Calculus for the Curious:
Copyright © 2020 by Kent Osband. All Rights Reserved.

No part of this publication may be reproduced, stored in a retrieval system or transmitted, in any form or by any means—electronic, mechanical, photocopying, recording or otherwise—without prior written permission from the publisher, except for the inclusion of brief quotations in a review.

For information about this title or to order other books and/or electronic media, contact the publisher:
Kent Osband
Mountain Brook, AL
www.risktick.com
info@risktick.com

Hardcover ISBN: 978-1-7343376-2-4   (size 6.14 x 9.25)
Hardcover ISBN: 979-8-7555444-1-2   (size 9.5 x 11)
Paperback ISBN: 978-1-7343376-3-1   (size 9.5 x 11)
Electronic ISBN: 978-1-7343376-4-8

Printed in the United States of America
Cover image by whiemocca/Shutterstock.com
Cover design by: Woven Red Author Services, www.WovenRed.ca

First published 11 May 2020
Last updated   12 April 2024

Contents

Acknowledgments — v

Chapter 0: Myths about Calculus — 1

Chapter 1: Nudges — 15

1.1 Rectangles — 15
1.2 Polygons — 20
1.3 Circles — 25
1.4 Notation — 30

Chapter 2: Sums of Nudges — 37

2.1 Areas — 37
2.2 Distance — 43
2.3 Integrals — 49
2.4 Fundamental Theorem — 54

Chapter 3: Slopes — 61

3.1 Tangent Lines — 61
3.2 Reflectors — 68
3.3 Four Core Rules — 76
3.4 Powers of x — 82

CONTENTS

Chapter 4: Slopes of Slopes — 87

- 4.1 Curviness — 87
- 4.2 Acceleration — 92
- 4.3 Centripetal Force — 97
- 4.4 Curve Design — 104

Chapter 5: Iteration — 111

- 5.1 Approximation — 111
- 5.2 Taylor Series — 117
- 5.3 Convergence — 122
- 5.4 Taylor Calculus — 129

Chapter 6: Logs and Antilogs — 136

- 6.1 Logarithms — 136
- 6.2 Antilogarithms — 144
- 6.3 Proportional Change — 148
- 6.4 Fluctuating Growth — 155

Chapter 7: Circle Functions — 163

- 7.1 Complex Numbers — 163
- 7.2 Complex Calculus — 170
- 7.3 (Co)sine Signals — 177
- 7.4 Tangents and More — 184

Credits — 189
Index — 195

Acknowledgments

This book aims to teach you calculus—or help you teach others—in the simplest, most intuitive, and most down-to-earth-and-up-to-the-stars way possible. To the extent I succeed, that is mostly to the credit of others. It is their ingenuity I am distilling, recounted in sources I did not compile, conveyed using tools I did not create through networks I do not manage. Some genius has so illuminated the field that we hallow it by name, like Archimedes, Newton, Leibniz, and Euler. But most of what I draw on reflect refinements over many generations by thousands of others, who figured out simpler ways to prove something or unified the treatment of seemingly disparate topics.

Learning things simply helps to declutter our minds. For an analogy, think of an aircraft carrier that keeps its flight deck clear. Simplified concepts give our imagination more room for takeoff and facilitate its landing with fresh insight.

For the sake of simplicity, this book is short. For the sake of intuition, it is filled with pictures. For the sake of practicality, it focuses on the problems that give rise to formulas rather than the formulas themselves.

I acknowledge that this won't be to everyone's liking. If you'd like more historical background and more thorough explanations, or just want to learn about calculus without having to learn it, a host of books do this better. My personal favorite is *Infinite Powers: How Calculus Reveals the Secrets of the Universe* by Steven Strogatz. He does a wonderful job explaining how wondrous calculus is.

Let me also acknowledge that this book isn't geared toward specific exam preparation. Most calculus exams require a lot of formula memorization or practice in solving specific types of problems quickly. This book focuses more on teaching a way of thinking that will let you rederive the formula you've forgotten or to pose a problem in a way that calculus can solve. Now that online resources like Wolfram Alpha can instantly remind you of formulas and grind out problem solutions, human memory speed matters less than ever and problem-posing skills matters more. But I am mindful that most students need to prove their mettle on exams. For them I recommend reading this book well before exam time and then shifting to problem-solving exercises.

I have written this book mostly on my own without the usual bevy of colleagues to thank for critiques. However, I did get substantial help near the finish line from three people. Joan Frantschuk designed the appealing cover while Shirley Osband and Elizabeth Dreher scanned for typos. I am very grateful for their assistance.

0

Myths about Calculus

Calculus offers some of the greatest problem-solving methods ever discovered. Anyone who understands basic algebra and geometry can learn it. The main challenge is the triple-D way calculus is usually taught: dry, dull and daunting.

Why Calculus Matters

Math has become humanity's most important toolkit, its most sophisticated common language. That is why both universities and employers prize math skills. The better those skills, the more easily you can tap the sea of human knowledge, and the more easily you can contribute to it. Calculus is particularly useful. Most majors in natural sciences, engineering, computer science and business administration require calculus, often as a prerequisite.

The name calculus comes from the Latin word for "tiny pebble". It decomposes big rocks of problems into tiny pebbles and searches them for insights. It helps us see interconnections. It helps us appreciate the wonders of the universe and the ways our tiny efforts contribute to a greater whole.

Unfortunately, the way most schools teach calculus is horrid. Too much focus on terminology and rules, not enough on comfort with the concepts. Too much memorization, not enough inspiration. Too many needless preliminaries, too few interesting results. It numbs minds and squeezes out the joy.

Joy? Calculus is supposed to be joyful? Yes, like music or sports. If you don't enjoy playing, you won't fully develop your talents or inspire others. Joy lets you experiment and fail without feeling a failure. Joy spurs you to practice to perfection. It's fun to figure things out!

Calculus can also help us communicate with the new form of life called artificial intelligence or AI. Machines use calculus-type methods to solve problems and communicate with each other. To understand them better and take better advantage of their skills, we need to learn how they think.

Six Myths That Look True

Like arithmetic or sports, calculus can't be mastered without a lot of thoughtful practice. Good teachers and coaches speed learning by simplifying new material and by cultivating intuition. Unlike arithmetic or sports, calculus is typically taught in needlessly complex ways with little emphasis on intuition. This reflects six myths about calculus:

- Kids' minds can't handle it.
- Requires precalculus first.
- Requires a lot of memorization and tedious computation.
- Irrelevant to coding.
- Devoid of social or spiritual interest.
- Inherently dry, boring, and hard to teach.

Widespread belief in these myths makes them look true. We don't give kids a chance to learn calculus simply. Standard curricula insist on teaching logarithms, exponentials, trigonometry and infinite series before they introduce calculus, using the name "precalculus" to sanctify it. Standard tests require a lot of memorization and tedious computation. We don't expose students to the use of calculus-type methods in computer algorithms. We don't relate the core notion of diverse, locally driven change to tolerant, entrepreneurial society or to the majesty of nature. We present the basics in stark, solemn and stilted forms.

Let's not confuse "can't" with "haven't tried". College-bound UK students typically get introduced to calculus two years before their US counterparts. Computer classes in the US teach kids object-oriented programming although a few decades ago this was considered ultra-advanced. It is wrong to treat math curricula as fixed roads with fixed speed limits. To view calculus afresh, we need to dispel these myths.

Myth #1: Kids' Minds Can't Handle It

Calculus uses tiny nudges to help analyze the whole. One can demonstrate the basics using playdoh and string, with hardly an equation in sight. Granted, calculus nudges are hard to grasp. Since they are inventions of our minds, nothing keeps us from subdividing them over and over, making them vanishingly small. Yet they still sum to something big and can approach stable ratios. Imagining infinite progressions, coupled with the strange notation calculus uses to identify these operations, can't help but induce momentary panic.

Don't worry. You—or your students if you're a teacher--have managed before. You learned to imagine numbers separate from the things you counted. You learned to imagine infinity. Somehow you got comfortable enough applying the concepts that you stopped worrying about their intangibility. You or your students can get comfortable with calculus the same way.

The only thing strange about this approach to calculus is its rarity. When a four-year-old points to the number 2, nobody tells her that can't be done, that no number is tangible, and that what she thinks is two is merely a cipher for something that's in common between sets of two bears, two pencils or two balls. Yet introductory calculus textbooks regularly demand more rigor than beginners can tolerate. What a waste of human sparkle.

Where is the number two?

Myth #2: Need Precalculus First

Most calculus was developed without what we know today as precalculus. Calculus-type methods were used to calculate the volume of a sphere over a thousand years before advanced algebra, trigonometry, logarithms, exponentials, infinite series or imaginary numbers were invented. All the core results of modern calculus were discovered more than a century before modern limit theory was developed, which is where most calculus textbooks start now.

In fact, calculus makes most precalculus easier to understand. It helps us graph functions and find their roots and extremes. It helps us appreciate the reflective properties of parabolas, ellipses, and hyperbolas. Exponential growth and e are barely comprehensible without calculus. Trigonometric functions are easy-to-graph components of complex exponentials. Trigonometric sums and differences follow immediately from exponential calculus. Convergence to limits makes more sense once we get comfortable using calculus.

If teachers want to introduce these topics early, great. Yet there is no need to make them prerequisites for calculus. It's like ordering budding basketball players to spend two years dribbling without taking a shot. I doubt that even slow learners benefit. Perhaps they're just bored stiff.

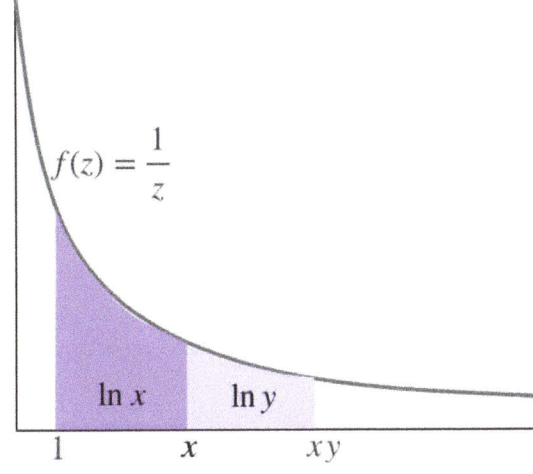

Presenting logarithms as areas under a c/x curve shows how they convert multiplication into addition.

Myth #3: Lots of Drudge

Grasping the concepts of calculus requires intuition, not computation. The only intimidating part is the notation.

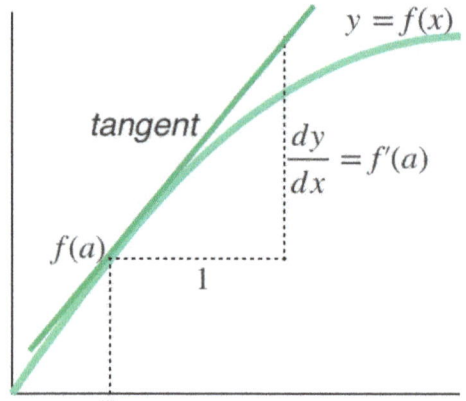

The derivative of a function indicates the rate of change and matches the slope of the tangent line.

The integral indicates an area under the curve and is the inverse of the derivative.

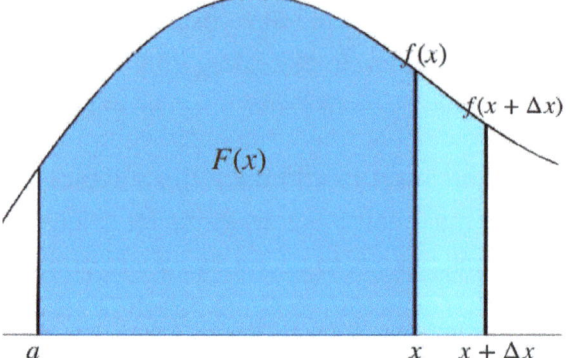

Calculation is straightforward and rests on four simple, easily-proved rules. There are fewer results to memorize than in arithmetic and fewer solution tricks to learn than in geometry. Most calculations are easily done online and there are a host of interesting applications, including:

- surface areas and volumes
- gravity, rockets and roller coasters
- optimization and curve-fitting
- half-lives, population growth and compound interest
- reflection and refraction of light
- springs and orbits.

Myth #4: Irrelevant to Computer Programming

Beyond addition and subtraction, most electronic computations embed calculus-type techniques. Searching through massive tables of stored results would be much slower. Calculus has inspired great ways to solve equations, create animations, and forecast user likes.

One very effective way to teach calculus is to have students code calculus operations into computer programs or spreadsheets. This helps students appreciate the usefulness of tiny nudges without being overwhelmed by boring iterations. A classic example involves estimating $\sqrt{x}$ by repeatedly averaging an estimate y with x/y to form a new estimate. Programming that into the computer builds more useful skills than calculating a square root by hand.

Myth #5: No Social or Spiritual Depth

Newton's calculus-driven discoveries spurred the Enlightenment. They revealed connections that transcended previous imagination. They raised objective reason and experimental confirmation above blind faith and sovereign decree. They sparked controversy and inspired poetry.

The main lesson of calculus is that small local interactions can mount into huge global or universal phenomena. This theme courses through physics, chemistry and biology, which is why they demand acquaintance with calculus. It is also important in economics, where it helps explain pricing behavior. Continual small adjustments help decentralized markets operate more efficiently and innovatively than central planning.

Myth #6: Dry and Boring

Calculus is usually taught in dry and boring ways because that's how teachers learned it. This book will refresh your sparkle by showing how fascinating calculus is and how easy it is to apply. Here are a few puzzles we will meet in subsequent chapters and use calculus to help solve:

CHAPTER 0

- Why are bubbles round?

- How does Earth manage to orbit the Sun without falling into it?

- Why do basketballs, bullets and jets of water move in parabolas?

- How do computers help us draw curves?

- Why does compound interest mount so fast?

- What makes springs oscillate?

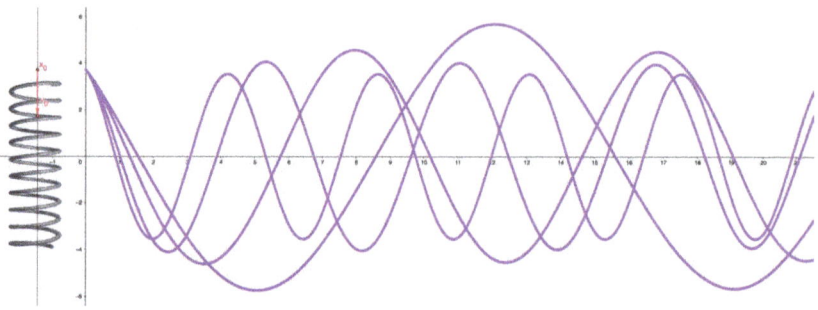

MYTHS ABOUT CALCULUS

- What do the necks of horses and the loops of roller coasters have in common?

- Why do racetracks have banked turns?

- Why do diamonds sparkle?

- How do eyeglasses work?

Approach

This book combines conventional pre-calculus and introductory calculus into a single course. To make this feasible, it reorders the topics, offers far simpler explanations, and sweeps away a host of clutter. In particular,

- It uses practical problems for motivation.
- It provides visual demonstrations of core concepts.
- It emphasizes general intuition over fine detail.
- It uses formal limit notation sparingly.
- It introduces differentiation and integration in tandem.
- It strips down derivations to application of four simple rules.

- It discourages rote memorization.
- It uses calculus to introduce iteration and infinite series.
- It uses calculus to introduce logarithms and exponentials.
- It uses calculus to introduce trigonometry.

I cover this material in seven chapters, each with four short 0arts and numerous illustrations. This is likely the shortest book on intermediate math you will ever see, the most colorful, and the least formal. I hope it is the most fun. As noted earlier, it presumes no more knowledge than basic algebra and geometry. Yet it covers 90% of the material on Calculus AP exams and tackles a few problems that go beyond.

The book moves fast if you let it and mostly you should. If there are just a few things you don't understand, don't worry about them. Maybe they're not that important, or maybe the next few pages will clear them up. Savor what you've learned without beating yourself up over what you haven't yet learned. For more background and detail, search online. Wikipedia at www.wikipedia.org nearly always provides a good survey.

I try to keep derivations brief. If you find them too brief, Wolfram Alpha at www.wolframalpha.com can solve nearly any calculus problem you're likely to see and will let you focus on particular parts. Its Pro Subscription provides detailed step-by-step solutions.

Should you ever find yourself thoroughly baffled, stop. Stop flipping pages forward and start flipping pages backward. Once you reach familiar territory, start reading forward again.

Prerequisites

Every game has prerequisites. Want to play water polo? First, learn how to swim. To play calculus, there are two things you need to learn first. One is basic algebra, which does arithmetic with letters. The other is basic geometry, which relates numbers to shapes. If you cannot graph simple functions—a task that combines basic algebra and geometry—you're not ready.

Not quite ready.

Expand $(a+b)^n$:

$(a + b)^n$

$(a + b)^n$

$(a + b)^n$

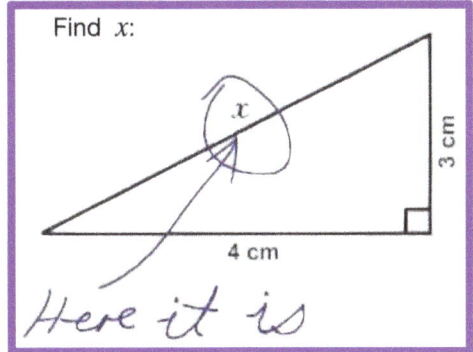

Find x:

Here it is

What else do you need to learn first? Not much. Just as you don't need to be a great swimmer to start water polo, you don't need to be great at algebra or geometry to start calculus. Playing calculus will help you improve more basic skills. Calculus also helps connect algebra with geometry, which makes both subjects easier.

While love of learning is vital, that depends far more on attitude than aptitude. Don't worry about your age either. While it is well-known that you can't teach an old dog new tricks, I interpret that optimistically. Keep learning new tricks so as not to become an old dog.

Accompaniments

To help yourself or your students get a tangible feel for calculus, play with modeling clay. Make curving snakes and trace their slopes. Make squares, triangles, circles, cubes, pyramids and spheres. Carve them into thin slices, measure the slices, and figure out what the slices tell you about the whole.

For more sophisticated demos, tap the vast online repository at www.geogebra.org. I have adapted over two dozen for this book. Search for "calculus" and you will find thousands more.

> *Over two dozen activities adapted for this book can be found by searching www.geogebra.org online for their name and/or the author tag "kosband".*

Like any other subject, no student can master calculus without a lot of practice and correction of mistakes. However, I haven't included any unsolved exercises apart from crossword puzzles. Why not?

- I want to keep this book short and self-contained. Since everyone learns differently, trying to cover everything will weigh us all down.

- There are many specialized exercise books and online apps that do this better. I am better at coaxing inspiration than perspiration.

- Error correction should be automated as much as possible, so that students get feedback sooner with less drudge for teachers. I have created a parallel online offering with exercises, videos, and instruction in calculus-related coding. Look for it on the web.

Outline

The rest of this book is organized as follows. Chapter 1 uses a simple problem—finding the maximum area for a given perimeter—to introduce the tiny pebbles of calculus. Section 1 focuses on rectangles, Section 2 allows general polygons, and Section 3 extends the argument to bubbles. Section 4 rephrases the discussion in terms of limits and derivatives.

Chapter 2 looks at sums of nudges. Section 1 uses them to calculate areas. Section 2 uses them to relate velocity to distance traveled. Section 3 introduces area functions, aka integrals. Section 4 presents the two Fundamental Theorems of Calculus relating derivatives to integrals.

Chapter 3 relates derivatives to slopes. Section 1 introduces tangent lines. Section 2 examines conic reflectors. Section 3 presents the four core rules of calculus. Section 4 applies these rules to find the derivative or integral of any polynomial of any power of a variable x.

Chapter 4 investigates slopes of slopes. Section 1 relates curve shapes to second derivatives. Section 2 applies Newton's Laws to analyze artillery fire and escape velocities. Section 3 analyzes centripetal forces. Section 4 examines curve design.

Chapter 5 applies calculus to potentially infinite series. Section 1 shows how to refine approximations through iteration. Section 2 introduces Taylor series that can approximate any smooth function. Section 3 investigates the convergence or divergence of Taylor series. Section 4 uses Taylor calculus to model constant relative growth, the integral of $1/x$, and oscillations of springs.

Chapter 6 looks at logarithms and antilogarithms. Section 1 shows how logarithms convert products and ratios into sums and differences. Section 2 demonstrates that antilogarithms are exponentials and relates them to interest compounding and radioactive decay. Section 3 applies exponential modeling to rocket needs, cooling, air drag, and present value. Section 4 analyzes fluctuating growth in damped oscillations, logistic growth, population pyramids and predator-prey models.

Chapter 7 relates trigonometry to circles. Section 1 identifies complex numbers with locations on a plane, identifies sines and cosines with coordinates of complex exponentials, and uses these identities to derive classic formulas for trigonometric sums and differences. Section 2 uses complex calculus to prove Kepler's three laws of planetary orbits.

Section 3 examines the refraction of light, the timekeeping of pendulums, and the integration of trigonometric polynomials. Section 4 introduces other trigonometric functions and applies them to integrate rational polynomials and analyze the spiraling flight of hawks.

A Vision of the Future

Let me present a vision of the future that I think is within our reach. I see countries in which most students learn basic algebra and geometry in middle school and get introduced to calculus by age 14. They will learn calculus intuitively and playfully, along the lines of this book. At least half of their schoolwork will entail individual and group projects exploring interesting applications of calculus. They will hone their skills less through direct calculations than through the code they write teaching novice computers to do the calculations for them.

No students will be expected to master calculus and pre-calculus in their introductory course, any more than they would a new sport or a foreign language. They will earn merit badges for the proficiencies they demonstrate and fill gaps later. As always, some students will advance faster or farther than others. But nearly everyone will learn the basics of calculus. No one will be considered to have a well-rounded education without it.

In the future, most students will clamor to take calculus because they heard it's so much fun. And it will be fun, thanks to their inspired teachers. Teachers who share their sense of wonder at how much calculus explains about the world and how well. Teachers who stretch their imagination to the infinitesimally small and the infinitely large. Teachers who guide them through the mysterious notation and help them unravel its meaning. Teachers who encourage playful problem-solving and cultivate their curiosity.

To join those teachers of the future, you will need to learn simple ways to analyze complexity and explain it to others. Let's start now.

1

Nudges

Calculus tackles problems through repeated nudges of trial solutions. Since nature operates in similar ways, calculus often not only solves problems but also provides insight into why the solutions work.

1.1. RECTANGLES

Designing a Playpen

I need to build a playpen outside for a puppy. I have 14 meters of fence and want to enclose as big a rectangular area I can. How?

One way is trial and error. Suppose I set width 6 meters and length 3 meters. Since a rectangle has four sides and opposite sides match, total perimeter equals $6+3+6+3=18$ meters. Oops, I'm 4 meters short of fence. To close the gap, I can halve the width. Now the perimeter is 12 meters, or 2 meters less than the fence. That can't be optimal either.

Hence the perimeter should equal 14—let's drop the meters as understood. Width and length should sum to half of that, so if we call the width x, length equals $7-x$ and the area is $x(7-x)$. Let's look for a solution through trial and error.

CHAPTER 1

Width	Length	Area
0	7	0
2	5	10
4	3	12
6	1	6
3.72	3.28	12.20

The best whole number for x is 3 or 4, either of which generates an area of 12. However, the answer isn't a whole number.

While trial and error is useful, it can waste a lot of effort and rarely tells us when to stop. For clues, notice a symmetry. Switch width with length and the area stays the same. The only potentially unique maximum makes the playpen a square of sides 3.5 each.

Do solutions have to be unique? No, but often they are, and if symmetry suggests one then it's worth examining. Let's form a square, nudge it, and see what happens. Here nudge means a tiny change in width versus length, while keeping the perimeter the same. Does it make the area bigger or smaller?

Nudges are so special that that they merit special names. One favored name is the Greek letter delta, written δ. The Greek makes it special, the "d" sound starts "difference," the pointy top points to tiny, and the rounded bottom resembles zero. In truth it doesn't matter. As Shakespeare might say, a nudge by any other name would act as small.

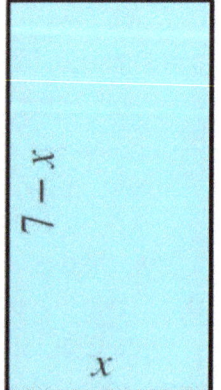
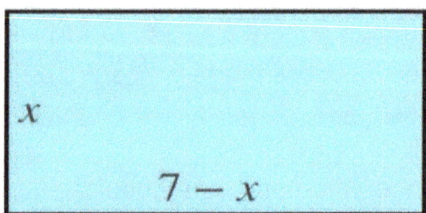

With a nudge δ pushing width to $3.5 + \delta$, length equals $3.5 - \delta$ for a total area of $(3.5 + \delta)(3.5 - \delta) = 12.25 - \delta^2$. Since δ^2 can't be negative, the area can never exceed 12.25 and the symmetric candidate with $\delta = 0$ is indeed best.

In this case, δ doesn't have to be small. We just need to notice the symmetry, guess the right answer, and check it. We won't always be so lucky or skillful. Might there be a way to get to the same answer even if our initial guess is wrong?

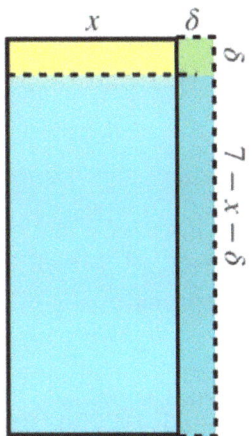

Yes, there is. Pick any starting width and nudge it by δ. Only this time make sure δ is truly small, no wider than a tiny pebble. All else being equal, this adds a long sliver of area $(7 - x)\delta$. Only all else isn't equal. Remember, the length must shrink to $7 - x - \delta$ in order to keep the perimeter the same. This trims a wide sliver with area $x\delta$ and chops off a tiny corner—a sliver of a sliver—with area δ^2.

The net change in area is $(7 - 2x - \delta)\delta$, or $7 - 2x - \delta$ per tiny pebble width δ. Round this to $7 - 2x = (7 - x) - x$, which equals length minus width. If length exceeds width, extra width boosts area. If width exceeds length, extra width reduces area. Area is maximized when width equals length. Is that true even when the perimeter isn't 14 meters? Definitely. The best rectangle is still square; all we're changing is the scale.

Calculus

The method of nudges helps guide its own trials and errors. From wherever you start, nudge width and length pebble by pebble until the area can't get any bigger. Often this reveals simplifying patterns—here, the importance of length minus width—that lead us sooner to the answer and help prove that it's right.

CHAPTER 1

Don't confuse simplification with simple-minded. It took genius to invent the number zero and the decimal system, but it made arithmetic vastly simpler. For example, in Roman numerals, VI·IX=LIV doesn't clearly imply LX·XC=vCD, but $6 \cdot 9 = 54$ does clearly imply that $60 \cdot 90 = 5400$.

You don't need to design the method of nudges to use it, any more than you need to design the decimal system. You just need to understand it and practice applying it. If you know basic algebra and geometry, are willing to think in new ways, and take joy in solving puzzles, you can learn the essentials in a few months. That is a wonderful privilege. Earlier generations had it much tougher.

The method of nudges is better known as calculus, which as noted earlier is the Latin word for "tiny pebble". Calculus breaks down big mental stones into countless tiny pebbles and sifts them for insights. However, the pebble width δ is weird. It can't be zero as that prevents comparisons. Yet any difference from zero can cause trouble.

For example, given a perimeter of 14, suppose current $x = 3.49$ when $\delta = 0.03$. Since length exceeds width, our decision rule accepts the nudge. Yet this doubles distance from the optimum because the nudge is too coarse. No matter how small we make δ, our decision rule can bog down very close to the optimum x. Call it a swamp or dead zone. To complete the solution, we need to drain the swamp.

The simplest drain treats δ as infinitesimal: a nonzero number so small we can barely distinguish it from zero. Through most of our investigation, we emphasize the non-zero aspects of δ but at the end we let it disappear. That is a fuzzy approach and can get us into trouble. Yet it works reasonably well in most cases and inspired most of the great discoveries in calculus. While I will introduce a more rigorous approach in a few pages, beginners shouldn't worry. You learned to talk without advanced theory. You learned to count without advanced theory. You can learn calculus without advanced theory.

Top of the Hill

For more insight, try a hiking game called "Top of the Hill." Go to a forest you've never seen before and see how quickly you can climb to the highest point, without using a map or asking for directions. To mimic the method of nudges, always move uphill. Stop when all the paths head out flat or down.

Will this work? With real-life hills, hardly ever. They have too many folds and furrows. If you're never willing to step down, you're bound to get stuck. Imagine a smart ant playing this game. It will halt on top of a pebble you didn't realize was there. To look smarter than the ant, it is better to ignore tiny ups and downs. If you always move toward the highest point in a 100-meter radius, you'll almost surely get to the top of a human-sized hill. Only that hill won't necessarily be the highest hill. For example, no nudges at the bottom of a steep canyon will tell you which side rises highest.

Why then do nudges work for our fence design problem? They work because our problem resembles a smooth hill with a single peak. To see this, let's fix a perimeter P and plot the area $x(\frac{1}{2}P - x)$ as a function of x. Along the horizontal axis, x varies from 0 to $\frac{1}{2}P$. Along the vertical axis, area increases gradually from 0 to $P^2/16$ and decreases gradually back to 0.

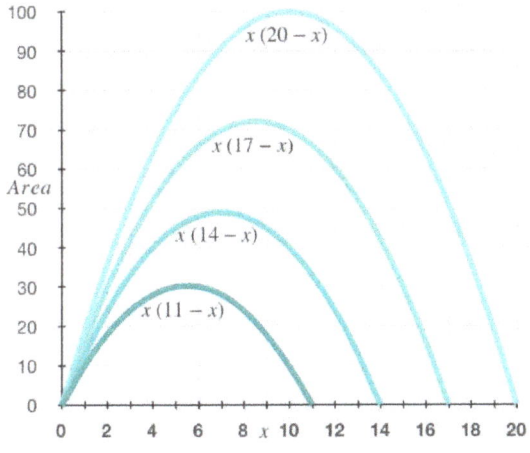

A microscopic hiker on any of these paths, nudging its way to the top, would act like the designer nudging the dimensions of the fence. That is the core connection between these two problems.

CHAPTER 1

The general problem type is known as optimization. It assigns a value to each possible choice and seeks the choice yielding the highest value. Provided the graph resembles a smooth hill, the method of nudges can help us find the top. The more learners practice this, the more comfortable they'll get with the connections.

1.2. POLYGONS

Triangular Fences

We have seen that the best rectangle for puppy's playpen is square. Only why insist on a rectangle? Let's use the method of nudges to compare a broader variety of shapes.

Consider a triangle, which we can form in two stages. First, pick two vertices and stretch fence between them to form a base. Second, pick the remaining vertex and extend fence around it to form a triangle.

Since triangle area equals half the base times height h, Stage 2 should aim to maximize h using whatever fence length L remains from Stage 1. Of the two sides it has to cover, let's call one length x, which makes the other $L-x$.

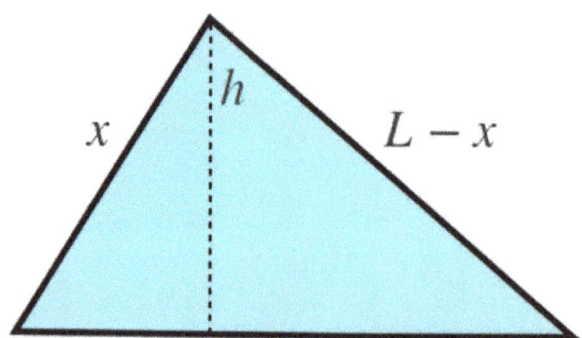

The symmetry of the problem encourages us again to guess that the two sides must be equal, forming an isosceles triangle. To confirm this, imagine a mirror laid face down across the top of the triangle and parallel to the base. The mirror image of the right side maintains its length $L-x$ and height h. The path from bottom left to upper right has total length L and height $2h$. To maximize h, the path must be straight.

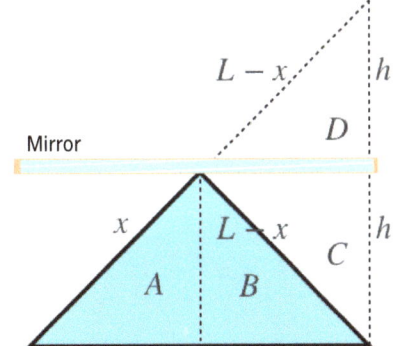

Here is an optimal triangle and its reflection. Triangles B, C, and D are mirror images. Triangles A and D must be congruent for their hypotenuses to form a straight line and their heights to match. Hence $x = L - x$.

By symmetry, any adjacent sides must match, so Stage 1 should make the base one-third of the perimeter, while Stage 2 should form an equilateral triangle.

Alternate Perspective

Another neat way to confirm that the sides must be equal involves nudges. If we stretch the left side to $x + \delta$ and swing the right side out to meet it, we add a triangular sliver of base δ and length $L - x$. To keep the perimeter the same we must shrink the right side by nearly δ, which

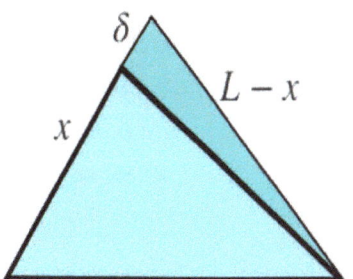
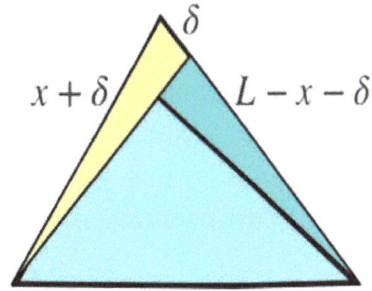

shaves off a triangular sliver with area close to $\frac{1}{2}\delta x$. The net change in area per tiny δ is close to $\frac{1}{2}(L-2x)$. Equalizing adjacent sides is best.

Note how simple this is. We don't need the ingenuity to drop a perpendicular, mirror a side, or construct a chain of congruent triangles. We just start somewhere and try to improve. We track the main slivers and dismiss the slivers of slivers. Can algebra alone solve this problem, without using nudges? Yes, but it's terribly messy and not very intuitive.

Isosceles Insights

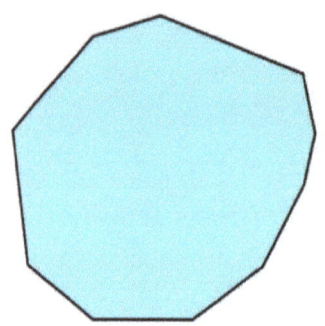
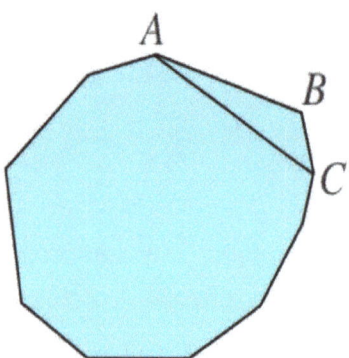

The equality of adjacent sides applies to any optimally sized polygon. For example, consider the decagon above. Pick any vertex B and draw a diagonal between its neighbors A and C. If we shift B without changing the sum of lengths AB and BC, neither the total perimeter nor the area below AC changes. All that changes is the area of triangle ABC. To maximize it, apply the isosceles insight and make AB and BC equal. If we keep equating adjacent sides, all sides will be equal at the optimum.

The isosceles insight also shows that more sides can beat less. Look again at the triangular slice ABC after we equate the lengths of AB and BC. Pick an arbitrary point D on AB as an extra vertex. Since side DB is shorter than side BC, we can shift B to increase area without changing the perimeter.

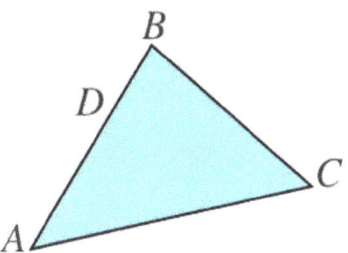 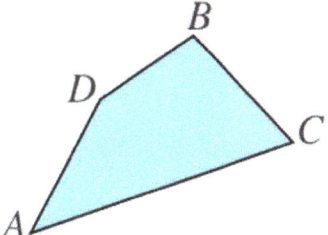

Area doesn't depend only on number and length of sides. For example, a quadrilateral with equal sides forms a diamond shape that might be squashed. Here are two ways to see that the optimal diamond for our fence is square:

- A diamond is a parallelogram, with area equal to base times height. Given the base, height is maximized by making the slanted sides stand straight up.

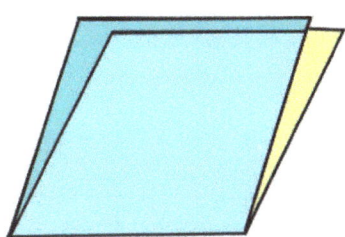 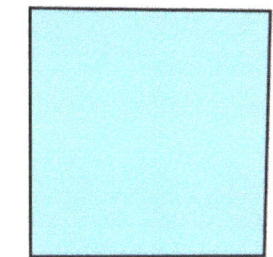

- Drop a perpendicular, chop off the protruding triangle, and move it to the other side. Since we'll enclose the same area with less fence, the slant couldn't have been optimal.

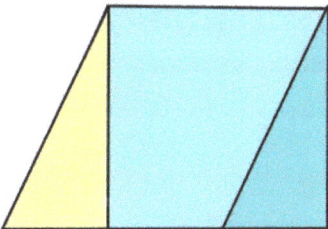

Regularity

By now you've likely guessed the answer. The best polygon should be "regular": maximally symmetric, with equal sides and equal angles. How can we demonstrate it? One method slices off a quadrilateral $ABCD$ and drops perpendiculars from B and C down to the unfenced side AD to

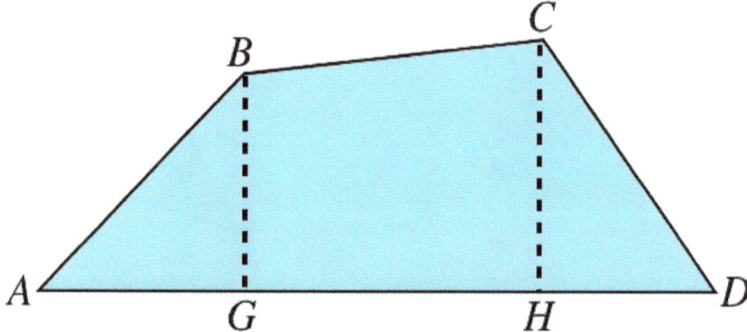

form right angles at G and H. When all sides have equal length, the interior angles will be equal if and only if BC is parallel to AD, since that makes the left and right triangles congruent.

Now that we know what we want, how do we get it? Suppose we rotate BC to raise the shorter end B by δ and lower C by nearly the same. The area of the quadrilateral $BCHG$ doesn't change. The shorter triangle ABG gains an area of about $\frac{1}{2}\delta$ times its width AG. The taller triangle DCH loses an area of about $\frac{1}{2}\delta$ times its width HD. Since $AG > HD$ (otherwise $AB < CD$, when they are supposed to be equal), the area grows. Ta-da. Making BC more parallel helps.

Again we see the merits of nudges. However, my presentation is sloppy. When BC rotates, it pulls on AB and pushes on CD. To keep those lengths the same, both G and H will shift slightly to the left, making AG a bit shorter and HD a bit longer. Fortunately, the differences in areas amount to slivers of slivers that we can ignore.

Are you feeling a bit lost in intersections and angles? Don't worry. Geometry experts needed centuries to solve this problem. All that matters for our purposes is gaining an intuitive feel for why:

- adjacent sides should be equal.
- adding extra sides can expand area.
- squashing the angles between sides tends to reduce area.

To strengthen your intuition, try the GeoGebra activity "Maximum Area with Fixed Perimeter". See how changing the number of sides affects the area.

1.3. CIRCLES

Round Design

In expanding the area a given length of fence can enclose, we have come to two main conclusions. First, set the sides in the shape of a regular polygon. Second, add sides.

Since it takes time, effort and posts to fasten a fence, and since each new side adds less area than the one before, at some point we say good enough and stop. Where this is heading is clear. Ignoring the costs of construction, the ideal solution is a regular polygon with an infinite number of sides and with each vertex equidistant from the center.

The process resembles a jeweler making a bracelet. Bend a rope of metal into a loop and fuse the ends. Gently hammer out kinks. Rotate the bracelet, un-squash whatever's squashed, and repeat.

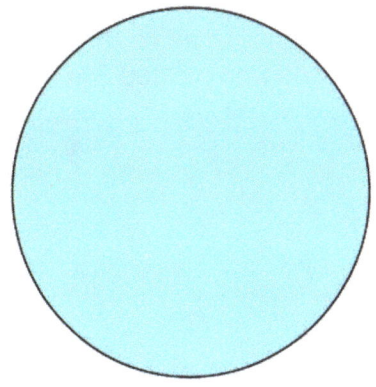

Amazing, that looks just like a circle. How does adding corners and flat edges get rid of both? The answer is: little by little. Each nudge makes corners slightly blunter and edges slightly flatter.

CHAPTER 1

Is the end result for a regular polygon truly a circle or just something infinitesimally close? Or is it unclear since the differences get tinier than we can possibly measure? Ah yes, the great nudge debate, this time in geometric form. To respectfully avoid it, let's redefine circles to include all lookalikes, but not pathetically lame imitations like the one below.

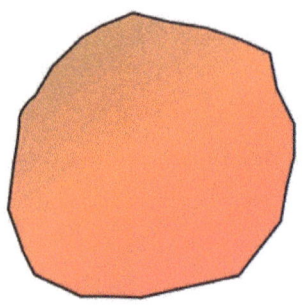

If my freehand sketches knew they were masquerading as circles, they would flush red with embarrassment. This one just did. Fortunately, modern technology devises vastly superior circles. It embeds them in wheels, basketballs, pizzas and other objects of delight. Computers let even miserable artists like me create circles to heart's content.

Our early ancestors lacked these advantages. Their only way to design a circle in one go was to fasten down one end of a rod and to mark the other end while it rotated. As for 3D circles, better known as spheres, no way. While rotating a flat circle around its axis can make a fine outline, the outline can't hold in place without noticeable corners, edges or flat faces.

In contrast, nature generates nearly perfect circles without any obvious scaffolding. This awed the ancients. They glimpsed something that transcended human ingenuity. At first they thought that big circles like Sun and Moon must be gods. Once people learned—after many generations of observation and analysis—to forecast solar and lunar behavior, they marveled less about Sun and Moon than about how they got constructed. The Sun god gave way to the god who created Sun.

Perhaps you think that's silly. Why get all bent out of shape about something that bends into shape? Yet if you don't see anything wondrous about circles then you are missing something important.

Bubbles

Tiny circles are at least as mysterious as big ones. Bubbles, for example, are transient and easy to destroy, yet far outnumber living things. Something must help them regenerate circles. What could it be?

Bubbles are gas in liquid skins. At the boundaries, the molecules of liquid attract each other as if infused with very weak glue. The stickiness bends surfaces slightly inward. When a surface wraps completely around a clump of gas, it forms a bubble.

Despite its peaceful looks, a bubble is a battleground. The gas molecules inside constantly jostle about and press to expand the volume. Surface tension presses back and tries to shrink the bubble. The only thing they effectively agree on is to maximize volume for a given surface area, or equivalently, to minimize surface area for a given volume. Like maximization of area given perimeter, the solution makes the surface round, only in 3D instead of 2D.

The explanation is gloriously simple and lays no burdens on gods. No "essence of sphere" needs to smuggle in. No special tool needs to plot boundaries. We just need zillions of aimlessly darting particles, two competing pressures, and continual nudges toward truce.

The explanation also hints of bubbles' life cycle. When bubbles expand or merge, their volume grows faster than their surface area. For example, doubling the radius of a sphere multiplies its surface area by four but its volume by eight. This makes bubbles increasingly fragile as they get bigger. Eventually they burst.

Conversely, shaking a mix of air and water stirs a swarm of tiny bubbles. There are said to be 100 million bubbles in every bottle of champagne. Now that is something to drink to!

CHAPTER 1

Plumpness

Plump is defined as "having a full rounded shape." Applied to fruit, it suggests something ripe and juicy, like in this painting.

Applied to animals, it suggests something well-fed and fat, like these two potbellied pigs.

In either case, plump gives the sense of innards stuffed until they stretch and round the outer skin. Like a water balloon or sack of flour, the shape might be far from spherical. Packaging, gravity, and internal supports can change it. Yet shape is always rounder with stuffing than without, unless the object gets so stuffed it explodes.

Why always? Why aren't there two kinds of stuffables: one that gets rounder when stuffed and one that gets sharper-edged? The answer again is nudges. The tension between expanding volume and constraining surface favors round. A particularly awe-inspiring example involves the gestation of babies. When a woman gets pregnant, her abdomen swells and gets rounder. Only later in pregnancy she will feel and sometimes see sharper protrusions of feet or elbows. It is a telltale sign that the fetus inside is more than a sack of flour or ripening fruit.

Wrinkles

Do bubbles and babies bore? Poor you. Yet the concepts also apply to more adult topics, like wrinkles.

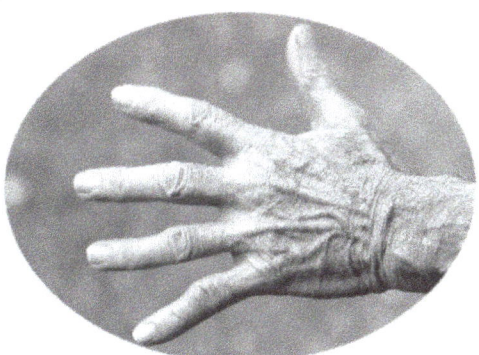

Skin wrinkles have two main causes. One is that skin gets less elastic as it ages and accumulates stretch marks and creases. Treatment demands a lot more than math skills, which explains why so few mathematicians moonlight as cosmetologists or plastic surgeons. The other main cause is loss of stuffing. When people lose a lot of weight quickly, loose skin can hang down in folds. Conversely, Mother Nature endows babies with extra folds of skin, to give them more to grow into. That's why newborn faces sometimes look very old.

The stuffing of most cosmetic concern lies in the skin itself. Its inner layers lose water and create wrinkles as they contract. Think of smooth grapes drying up into wrinkled raisins. The best antidote other than not wrinkling in the first place is rehydration. Raisins soaked in water plump up again. However, this doesn't work for people because the outer skin is defensive and won't let the water get through.

The trick then is to find substances that will smuggle the water through the outer layer of skin and get soaked up by the inner layer. They're called moisturizers. Add some fragrance, color, and silky texture, package it and offer it for sale. There are many ways to do this: Amazon lists over 60 thousand. Will calculus help you make a fortune selling moisturizers? I doubt it. Still, the moisturizer business and calculus share at least two core principles. First, big results start small. Second, tiny nudges can smooth.

CHAPTER 1

1.4. NOTATION

What's in a Name?

Among the challenges facing calculus beginners, notation is often the biggest. That's a shame. It's like naming a newborn imp Rumpelstiltskin. Constantly ridiculed, the sweet imp turns sour. He holds queens' daughters for ransom and stamps through floors in rage. Would it not be vastly better to name him Fred, perhaps with extra respect by writing in Greek letters: $\Phi\rho\varepsilon\delta$?

To avoid trauma with these next-to-nothing *calculi*, I give them friendly names like nudge and write them with Greek letters. Unfortunately, this doesn't take us far. The core problem is that nudges aren't stand-alone variables. They are actions or "operators" that transform other variables.

The best-known operator is negation or "minus". Negation converts any x into $-x$, where $x + (-x) = 0$. Operators can operate on each other, as in $-(-) = +$. However, operators usually appear attached to variables, as in $-(-x) = +x$, since this links actions with results.

Like negation, nudges can refer to either actions or results. Recall our initial problem: nudge a rectangle with fixed perimeter to expand its area. We solved it using just two symbolic names: x for width and δ for the nudge's impact on x. If we name multiple variables, it can be hard to recall which impact is associated with which fence property. How might operator notation help?

The Δx Convention

Analogy with negation suggests a straightforward solution. Let a symbol like Δ (the capitalized version of δ) denote the nudge operator and let Δx and Δy denote the impact on variables x and y respectively. We can describe Δx as "change in x" or "delta x". To apply to our rectangular fence problem with initial area $y = x(7-x)$, let $y + \Delta y = (x + \Delta x)(7 - x - \Delta x)$ denote the area after x is nudged to $x + \Delta x$. The ratio of change in y to change in x is $\dfrac{\Delta y}{\Delta x} = \dfrac{(y + \Delta y) - y}{\Delta x}$.

So far so good. The Δx replaces δ and reminds that x is being changed. The Δy highlights the impact on y. Their ratio tells us something useful. Unfortunately, it seems to beg the simplification $\Delta y / \Delta x = y/x$, which is usually terribly wrong.

Treating these ratios as equal will usually kill our attempted solution. Better to avoid it like poison.

Why does a simplification so wrong look so right? The problem is less the operator notation than the habit in algebra of dropping multiplication symbols • or ×. As a result, Δx and Δy look as if a Δ variable is multiplying both x and y. Since we want to simplify where we can, it pains not to cancel out the two Δ.

Adding to the confusion, canceling out two like operators isn't necessarily wrong. In the case of negation, $\dfrac{-y}{-x} = \dfrac{(-1) \cdot y}{(-1) \cdot x} = \dfrac{y}{x}$ for any y and nonzero x. In the case of Δ, $\dfrac{\Delta x}{\Delta x} = 1$ and that helps to simplify $\dfrac{\Delta y}{\Delta x}$. But what we cancel out are equal changes in x, not Δ by itself.

CHAPTER 1

Now that we've seen how to do the calculation wrong, let's try to do it right. For neater computation with less chance of error, switch Δx back to δ and substitute z for $7-x$.

For $z = 7-x$ and $\delta = \Delta x$,
$$\Delta y = (x+\delta)(z-\delta) - y$$
$$= xz - x\delta + \delta z - \delta^2 - y$$
$$= -x\delta + \delta z - \delta^2$$
$$\frac{\Delta y}{\Delta x} = \frac{\delta(-x+z-\delta)}{\delta}$$
$$= z - x - \delta = 7 - 2x - \Delta x$$

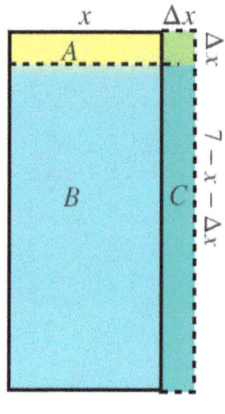

The best way to keep the calculations straight is to connect them with the geometry. In the picture to the left, y sums the areas of A and B while $y + \Delta y$ sums the areas of B and C, so Δy equals the area of C minus the area of A. Divide by Δx and we get the length of C minus the width of A, for the same answer as above.

Limits

Suppose we derive an expression like $\Delta y/\Delta x = 7 - 2x - \Delta x$. As we shrink Δx, the right-hand side gets closer and closer to $7 - 2x$. How might we say that more crisply?

It is tempting to substitute $\Delta x = 0$, since that gets the right-hand side where we want it. But then the left-hand side divides zero by zero. On reflection, what we most want is an infinitesimal nudge that is tinier than any fraction. This is known as the differential and prefixed with letter d. We can then write $\frac{dy}{dx} = 7 - 2x - dx$ and drop the last dx as insignificant.

That leaves two problems. First, dx looks even more like a multiple of x than Δx does. Second, the only real number tinier than any fraction is zero. Do differentials have reasonable hope of imitating real numbers or are they just hype about numbers that cannot be? Modern consensus lies somewhere between hope and hype.

An alternative that everyone accepts involves limits. Consider a sequence of Δx values that gets "ever-so-close" to zero, meaning that no nonzero number gets closer. Examples include the reciprocals of $1, 2, 3, 4, \ldots$, $1, -2, 3, -4, \ldots$, or $1, 2, 4, 8, \ldots$. If we feed the values one by one into the equation $\Delta y/\Delta x = 7 - 2x - \Delta x$, this gets ever-so-close to $7 - 2x$. Calculus phrases this as $\lim_{\Delta x \to 0} \frac{\Delta y}{\Delta x} = 7 - 2x$, which is read "limit of delta-y over delta-x as delta-x approaches zero is $7 - 2x$".

In ordinary speech, a close approach can stop short, back off, or run past. To make clear that Δx gets ever so close and stays there, it is better to say "converge to" than "approach." For more precision, let's call a set of numbers "δ-close" to a limit L if all values lie within δ of L. A sequence gets δ-close to L if it is δ-close except for some values at the beginning. To get infinitesimally close, it needs to get δ-close for any positive δ, no matter how small. If so, the sequence is said to converge to the limit L.

To say this formally, a sequence converges to L if and only if for every positive δ we can find an N such that every value after the first N lies within δ of L. While that sounds stilted, it strikes a balance between setting a clear destination and leaving ample freedom for how to get there. For the sequence $1/1, 1/2, 1/3, 1/4, \ldots$ any $N > 1/\delta$ will work. For the sequence $1/1, 1/2, 1/4, 1/8, \ldots$ any N satisfying $2^N > 1/\delta$ will work.

Graphically, a sequence converges when it enters a kind of funnel or vortex. The vortex squeezes everything in it toward a single point, although it might take forever to get there.

CHAPTER 1

Here are some sequences that converge along with their limits:

- $1, 0.1, 0.01, 0.001, 0.0001, 0.00001, \ldots \to 0$
- $1+1, 1.1, 1.01, 1.001, 1.0001, 1.00001, \ldots \to 1$
- $1, 1.1, 1.11, 1.111, 1.1111, 1.11111, \ldots \to 1\frac{1}{9}$
- $2.1, 1.9, 2.01, -1.99, 2.001, 1.999, \ldots \to 2$.

Here are some sequences that do not converge: $1, 2, 3, 4, 5, \ldots$, $1, -1, 1, -1, 1, -1, \ldots$, and $0, 1, 0, 0, 1, 0, 0, 0, 1, 0, 0, 0, 0, 1, \ldots$. A series that grows unboundedly high or unboundedly low is often described as getting infinite, with infinity written ∞ or $-\infty$ depending on the sign.

When sequences involve ratios, it can be hard to figure out whether they converge, much less what they converge to. The following procedure solves most problems:

- Factor out like terms to avoid division by 0 or multiplication by ∞ as in $\lim\limits_{x \to 0} \dfrac{3x^2 - 4x}{2x^2 + 3x} = \lim\limits_{x \to 0} \dfrac{3 - 4/x}{2 + 3/x} = \lim\limits_{x \to 0} \dfrac{3x - 4}{2x + 3}$.

- If numerator or denominator converge to nonzero constants, substitute the constants, as in $\lim\limits_{x \to 0} \dfrac{3x - 4}{2x + 3} = -\dfrac{4}{3}$.

- If numerator converges to 0 and denominator does not, the limit is 0, as in $\lim\limits_{x \to 0} \dfrac{3x}{2x + 3} = 0$.

- If denominator converges to 0 and numerator does not, there is no finite limit. Write ∞ or $-\infty$ if the numerator has a limiting sign, as in $\lim\limits_{x \to 0} \dfrac{3x - 4}{2x} = -\infty$.

- When $x \to \infty$ or $x \to -\infty$, substitute $z = 1/x$ and take the limit as $z \to 0$, as in $\lim\limits_{x \to \infty} \dfrac{3x - 4}{2x + 3} = \lim\limits_{z \to 0} \dfrac{3/z - 4}{2/z + 3} = \lim\limits_{x \to 0} \dfrac{3 - 4z}{2 + 3z} = \dfrac{3}{2}$.

- When both numerator and denominator converge to 0 despite our best efforts, apply L'Hôpital's Rule described in Chapter 5.

Differentials

We don't need to worry about differentials and their do-they-or-don't-they-exist dilemmas. However, differentials persist in the terminology calculus uses. The most common shorthand for $\lim_{\Delta x \to 0} \frac{\Delta y}{\Delta x}$ is $\frac{dy}{dx}$, just like the ratio of two differentials. This is known as Leibniz notation and is said to describe a "derivative".

Differentials also persist in our thinking about limits. It is much easier to imagine one maximally shrunken nudge than an ever-shrinking sequence of nudges. Granted, there's no such thing as maximally shrunk, so when something smaller comes along our nudge will have to shrink more. Potentially this can continue forever, which converts the differential back into the limit of an ever-shrinking sequence. Still, the differential form is easier for beginners to play with.

It took roughly two centuries of squabbling for most of the mathematics profession to convince itself that limits of converging sequences are the right way to think about calculus and that differentials are the wrong way. Yet the two approaches so closely resemble each other that we often use the same terminology to describe them with the same results.

Do you find that a bit silly? I hope so. It reminds of the long-standing controversy over who really wrote the wonderful plays credited to Shakespeare. Mark Twain joked that the true author was someone else with the same name. In that spirit let's agree that calculus doesn't use differentials but something that acts just the same.

Humor can assuage panic over differences in formulation or interpretation. However, there's no avoiding the challenge of notation. Just as algebraic variables need symbols other than ordinary numbers, these vanishing nearly nothings and their limiting ratios need symbols other than ordinary algebra. Learners need to get accustomed to the manner of expression. To review:

- The change Δx (read "delta-x") refers to a change or nudge in x.
- The differential dx (read "d-x"), refers to an infinitesimal Δx.
- The derivative $\frac{dy}{dx}$ (read "d-y-d-x") refers to $\lim\limits_{\Delta x \to 0} \frac{\Delta y}{\Delta x}$.

We will meet other types of calculus notation later. They reflect the multiple ways we can think about calculus concepts and apply them. Most importantly, derivatives have partners that, so to speak, work in reverse. They use limits too but in a very different way and have their own special expressions. The next chapter will introduce them.

2

Sums of Nudges

Sums of infinitesimal nudges can solve many problems. Called integrals, they correspond to areas of geometric figures. Nudges of these areas correspond to derivatives. The close link between integrals and derivatives is fundamental to calculus.

2.1. AREAS

Multiplication

Calculus has two main operations, distinguished by how they use nudges. Differentiation computes the ratios of nudges, called derivatives. Integration computes the cumulative sums of nudges, called integrals. Integrals are best viewed as the areas or volumes of shapes. To better appreciate how integration works, use modeling clay to imitate the procedures described here.

The simplest integration computes the area of a rectangle. To make this more interesting, imagine we wake one morning with "mathnesia": partial math amnesia. We recall addition, the real number line, and something about multiplication as repeated addition, but little else. How can we refresh our memory?

CHAPTER 2

Let's start by making a bunch of equal-sized squares. Lining up m of them in a row creates a rectangle with area m.

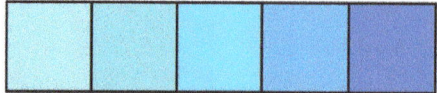

Aha, we see repeated addition, so $m = m \cdot 1$.

Stacking an identical row on top creates a rectangle with area $m + m$. That's just more addition, so $m \cdot 2 = m + m$.

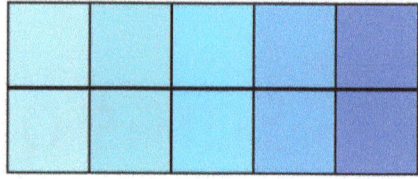

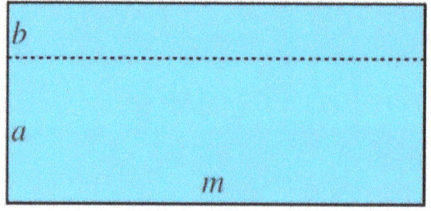

Before long we realize that $m(a+b) = m \cdot a + m \cdot b$ for any set of whole numbers. We have rediscovered the distributive rule!

Rotation shows that the order of multiplication doesn't matter any more than the order of addition does. That's the commutative rule.

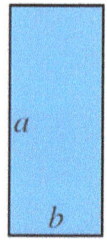

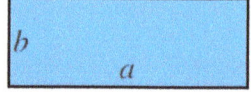

$5 \cdot 2 = 2 \cdot 5 = 5 + 5 = 10$
$5 \cdot 3 = 5 \cdot (2+1) = 5 \cdot 2 + 5 \cdot 1 = 15$
$5 \cdot 4 = 5 \cdot (3+1) = 5 \cdot 3 + 5 \cdot 1 = 20$

Using these rules, we can work out the product of any two whole numbers.

Next we divide a square into m equal strips. This is the visual representation of $m \cdot \frac{1}{m} = 1$. Redo the previous constructions with the smaller strips to see that $\frac{a}{m} \cdot b = \frac{a \cdot b}{m}$.

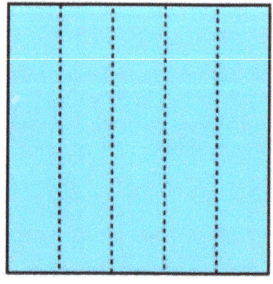

For irrational numbers, use limit arguments. Multiply fractions that are a bit too small, multiply fractions that are a bit too high, and squeeze them toward the answer between.

Triangles

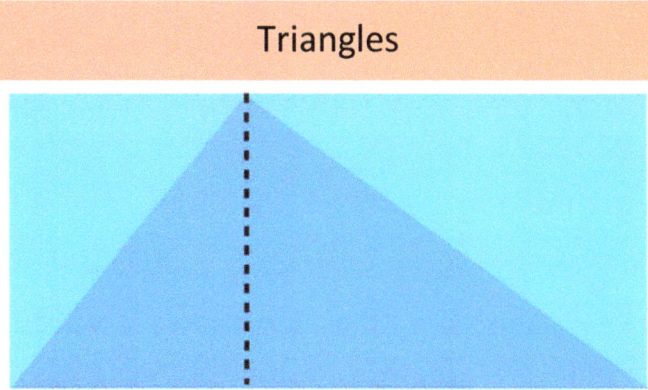

Having recalled the area of rectangles, let's tackle triangles. Every triangle can be split into two right triangles by dropping a perpendicular. Every right triangle can be joined with a complementary clone to form a rectangle. Hence the area of a triangle equals half its base times height.

To prove the Pythagorean Theorem, take any right triangle, clone it three times, and join them all head to toe. Do it in a way that makes both inner and outer boundaries form squares.

Letting a and b denote the triangle sides and c denote the hypotenuse, the smaller square has area c^2 while the larger square has area $(a+b)^2 = a^2 + 2ab + b^2$. The difference in areas must equal the area $2ab$ of four triangles, which implies $a^2 + b^2 = c^2$.

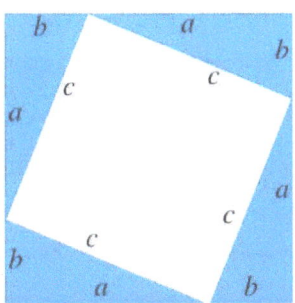

Suppose we're struck again with mathnesia, or perhaps never learned to play with triangles. Perhaps we don't like any angles other than right. Is there any way to compute the area of a solitary triangle using only rectangular strips?

At first glance there isn't. No rectangle can generate a slanted edge opposite a flat edge. For an approximation, take a bunch of rectangular strips, clip them at different heights, and use them to build a staircase. There's no slant anywhere, just flats and jumps. Still, the narrower the stairs, the more the staircase resembles a triangle.

CHAPTER 2

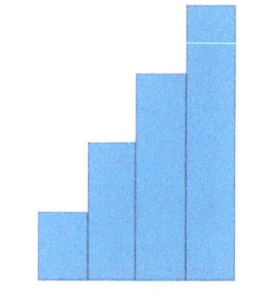 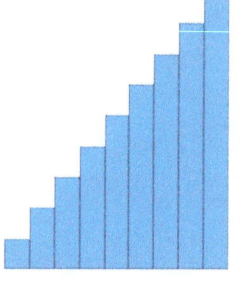 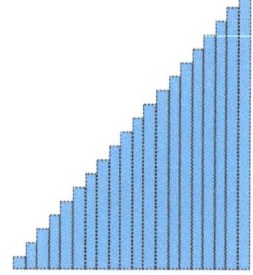

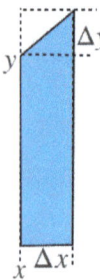

Every rectangular strip approximates a vertical slice of triangle over the same interval between x and $x+\Delta x$. It ought to include as much of the slice as it can and as little of anything else. Set the strip height somewhere between the heights y and $y+\Delta y$ at the edges. This bounds the strip area between $y \cdot \Delta x$ and $(y+\Delta y)\Delta x$.

To simplify the fine tuning, let us divide the whole triangle of width w and height h into n slices of equal width. Here are the dimensions of the best lower-bound strip and best upper-bound strip for each slice, numbered from smallest to largest.

Slice	Width	Lower-Bound Height	Upper-Bound Height
1	w/n	0	h/n
2	w/n	h/n	2h/n
3	w/n	2h/n	3h/n
...	...	...	...
n	w/n	(n-1)h/n	nh/n = h

Notice the pattern? Every upper-bound strip has a matching lower-bound partner one slice higher, except for the last upper-bound strip of height h. Hence total upper-bound area exceeds total lower-bound area by hw/n, which approaches 0 for huge n.

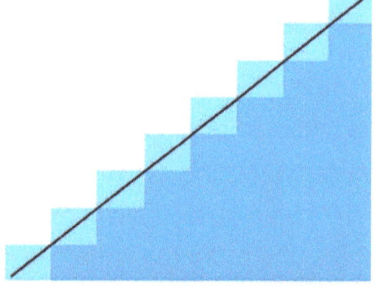

Narrowing the strips is core to integration. Here triangle area equals wh times $S(n) = (1 + 2 + \cdots + n)/n^2$. To confirm that $S(n) \to \frac{1}{2}$, pair off the numerator with its reverse sequence $n + \cdots + 2 + 1$. Since each of the n pairs sums to $n+1$, $2S(n) = n(n+1)/n^2 = 1 + 1/n$. To picture this, clone an n-step staircase, rotate it upside down, and stack it on the original.

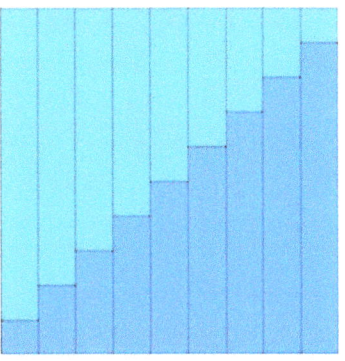

Circles

Having refreshed our triangle memory, let's move on to circles. Suppose we recall that the ratio of circumference to diameter goes by the name "pi", written π. Given radius r, circumference equals $2\pi r$. Unfortunately, mathnesia has garbled the formula for circle area into "pies are square," which can't be right. How might we recover πr^2?

The best-known approach starts with a regular n-sided polygon. Connect vertices to center to form n equal triangles. Since each triangle has area of half the base times height h, and since bases sum to perimeter P, total area equals $\frac{1}{2} Ph$. The largest regular polygon inside a circle has vertices that just touch the edge, with $P < 2\pi r$ and $h < r$. The smallest regular polygon surrounding a circle has sides that just touch the edge, with $P > 2\pi r$ and $h > r$. The difference in area between these polygons approaches zero as n grows. Hence the squeeze principle applies again with limiting area $\frac{1}{2}(2\pi r)r = \pi r^2$.

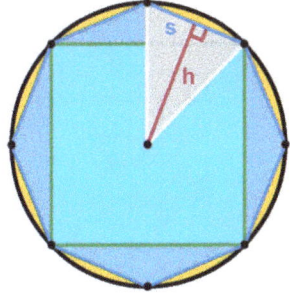
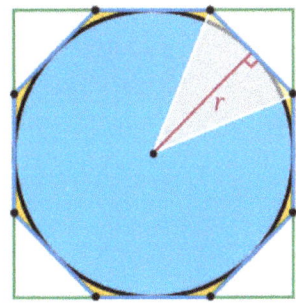

CHAPTER 2

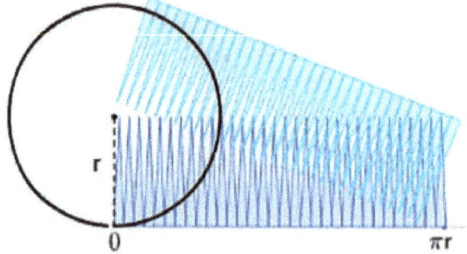

The GeoGebra activity "Area of Circle" makes the convergence more intuitive.

Another rearrangement uses circular strips. Imagine that the rings of this red onion are perfectly round. If we snip a ring from radius x to $x + \Delta x$ and lay it out flat, it will resemble a trapezoid of height Δx and width $2\pi x$ to $2\pi(x + \Delta x)$. Stack the flattened rings on top of each other to obtain a near-triangle of base $2\pi r$ and height r.

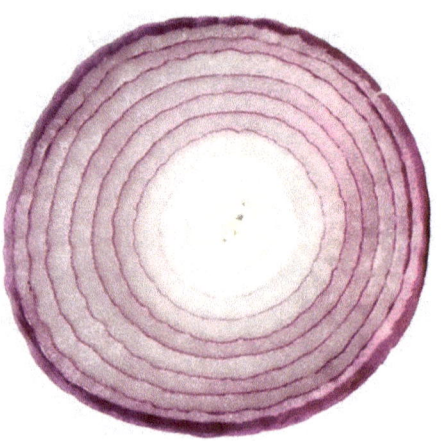

How about rectangular strips? In the half-circle below, a thin rectangle with base x away from the origin is approximately $\sqrt{r^2 - x^2}$ high. These strips are harder to sum than the rings or triangular wedges. However, the limiting value must be the $\frac{1}{2}\pi r^2$ area of the semicircle. This is a general principle. In calculating areas, we can use whatever method is easiest and be confident that other results must match.

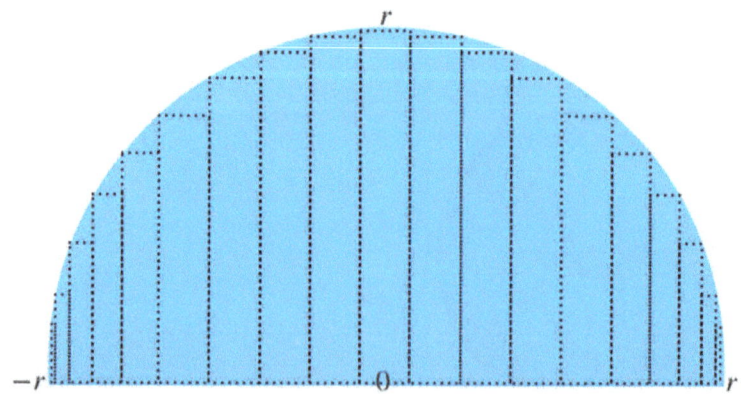

2.2. DISTANCE

Fetch

Last winter a pond near our house froze so solid that puppy and I played ball on it. Suppose puppy ran 4 meters per second while the ball rolled 2 meters per second. If the ball had a head start of 20 meters, how far did puppy run to catch the ball and how long until puppy brought it back?

The easiest solution to this problem takes puppy's point of view. Every second it gained 2 meters on the ball, so it closed the gap in $20/2 = 10$ seconds. In 10 seconds it ran $10 \cdot 4 = 40$ meters. If puppy turned around instantly (as we'll assume for simplicity) and ran back just as fast, it needed 20 seconds to complete the round trip.

Let's chart absolute velocity v as a function of time t. To recall, a function maps input values to output values. In this case, our function is simply $v = 4$ for $0 < t < 20$.

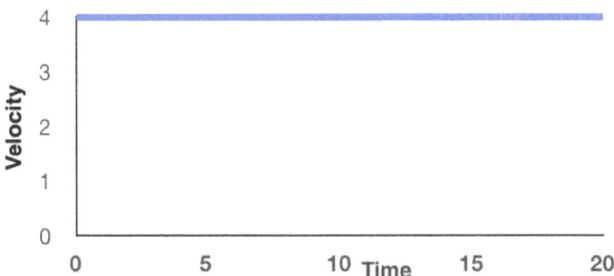

This horizontal line can easily sprout a rectangle underneath. Just add the two axes as boundaries and draw a vertical line at some t. The rectangle will be 4 high and t wide, with area $4t$. What does $4t$ tell us? How far puppy ran in all.

Suppose v doesn't equal 4. As long as it stays constant, puppy will run vt in t seconds. For any t, the area vt matches the distance run.

43

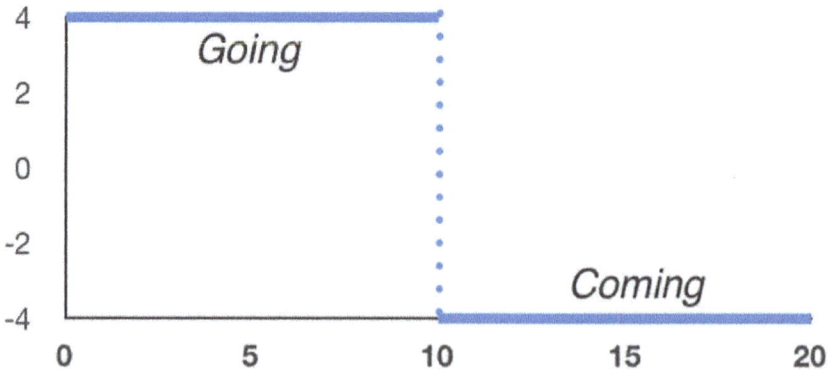

Another relevant measure is distance y from the thrower. By that metric, puppy's velocity coming back is -4 rather than $+4$. The revised graph has two distinct segments, with dots between them to note where velocity switches. How can we relate this to y?

The answer is: start every area strip from the horizontal axis. Going to the ball, the rectangle is the same as before. Coming back to thrower, a second rectangle extends below the axis. To calculate net distance, add areas above the axis and subtract areas below the axis, or equivalently, add the latter with negative signs to reflect their negative height.

Why does this work? Recall that velocity v equals the ratio of change Δy in location (displacement) to change Δt in time. The corresponding rectangle has area $v \Delta t = \Delta y$ if negative area is allowed and the absolute value $|\Delta y|$ if not. For areas to match displacements, their values must be signed.

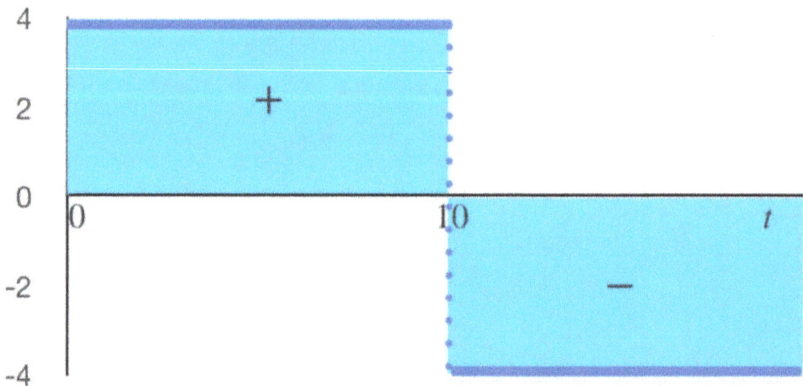

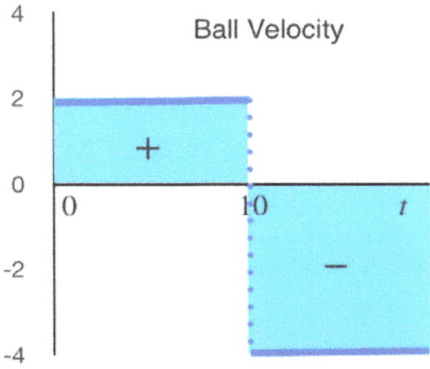

When we chart ball velocity, the net areas don't match net areas for puppy. Why not? Simple: Net area measures displacement rather than location. To convert to distance from thrower, we need to add the head start of 20, which isn't on the chart.

Sledding

Another activity puppy and I enjoyed last winter was sledding. Starting on a smooth slope, our sled gathered speed until landing in a snow-covered pile of leaves. Suppose our velocity v in meters per second equaled 0.8 times the number of seconds t. How far had we traveled in 4 seconds? How long did it take to reach the pile 40 meters away?

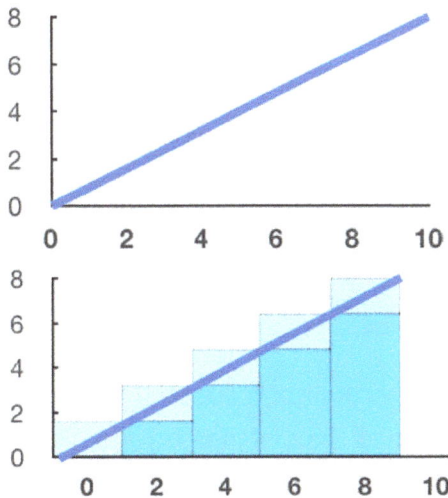

The chart of sled velocity is a diagonal straight line. We can approximate it by the steps of a regular staircase, like we did for triangles.

Two such staircases are charted to the left. One wedges its steps just under velocity. The other wedges its steps just over velocity.

The area of each staircase indicates how far the sled would travel if velocity alternately jumped and stabilized to match the steps. Hence the areas set lower and upper bounds for distance traveled. The narrower the steps, the closer the upper and lower bounds squeeze together toward the triangular area below the velocity line.

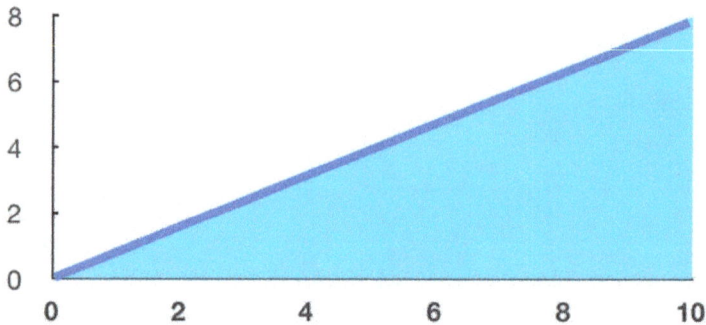

From here calculation is easy. At any time t, the base is t and the height is $0.8t$, for a triangle area of $0.4t^2$ that indicates the distance traveled. Hence it took $t=10$ seconds to travel 40 meters.

To generalize, let $v=at$ for some constant a. For a base that stretches from 0 to t, the height of the triangle is at, so both area and distance traveled equal $\frac{1}{2}at^2$. To generalize further, suppose our sled gets an initial push, or we don't start timing until we're already on our way. The revised equation for velocity is $v=at+b$ for some constant b. The velocity and underlying area are charted below for a and b positive.

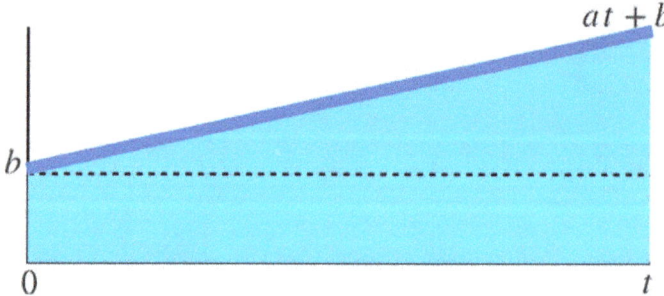

Again, the shape can be viewed as the limit of staircases, where the rectangle under each step approximates the distance traveled during that interval. Since the area of a trapezoid equals base t times average height $\frac{1}{2}at+b$, total distance traveled is $\frac{1}{2}at^2+bt$.

What if mathnesia strikes again and wipes out any memory of trapezoids? No problem. Draw a horizontal line at b to divide the shape into two: a triangle above and a rectangle below. Then calculate the areas separately and add them, for the same answer $\frac{1}{2}at^2+bt$.

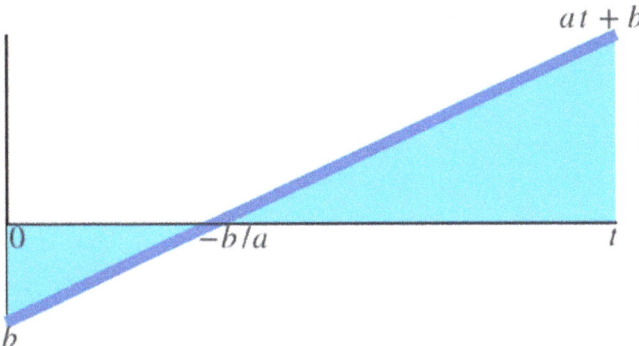

What if someone shoves our sled uphill before we slide back down? This corresponds to a negative b. Velocity reaches zero at time $-b/a$, which divides the area between velocity and horizontal t–axis into two triangles. As discussed earlier, treat area below the axis as negative. The net area works out to $\frac{1}{2}at^2 + bt$, so the formula stays the same despite the change in sign of b.

Displacement

Given any velocity function and starting time, the displacement is also a function. It maps each time t to the net distance traveled since the start. Now that we have examined various velocity functions, let's graph the associated displacement functions, assuming a start at $t = 0$. The simplest case is constant velocity b. Displacement $y = bt$ is a straight line through the origin. For puppy running to ball and back with constant absolute velocity, displacement outlines an isosceles triangle.

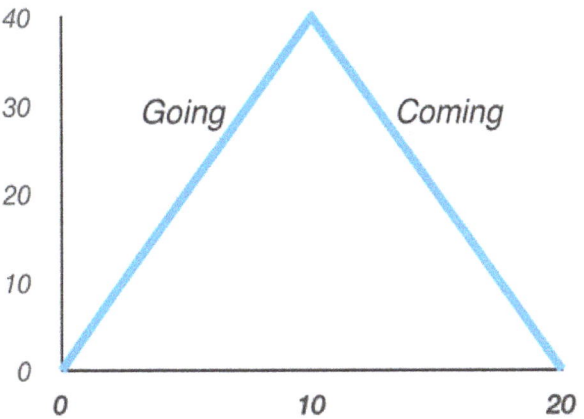

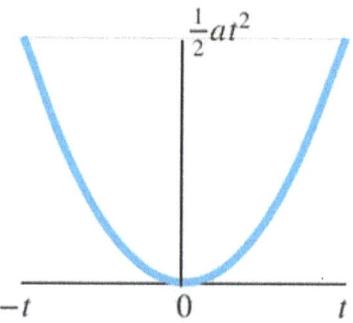 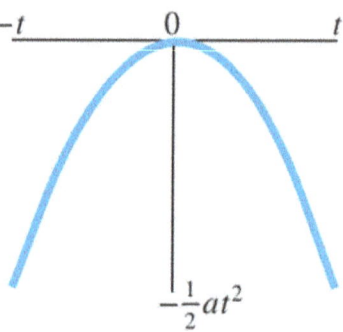

When $v = at$, displacement $y = \frac{1}{2}at^2$ is a quadratic function that is symmetric around the vertical axis and reaches a maximum or minimum at the origin. The curve bends upward for $a > 0$ and downward for $a < 0$, in a shape known as a parabola.

When $v = at + b$, the displacement . $y = \frac{1}{2}a t^2 + b$. still has a parabolic shape but the minimum or maximum shifts. The new axis of symmetry is the time $t = -b/a$ when $v = 0$ and $y = -b^2/(2a)$. Intuitively, $v = 0$ is where direction reverses so this is bound to mark an extreme.

In short, a constant rate of change in velocity is associated with linear velocity and parabolic displacement. If you knew that back in 1600, barely four centuries ago, you could have teamed up with Galileo Galilei, a father of the scientific revolution. He was the first to recognize the connection and use it to solve practical problems. We'll use it to solve practical problems too.

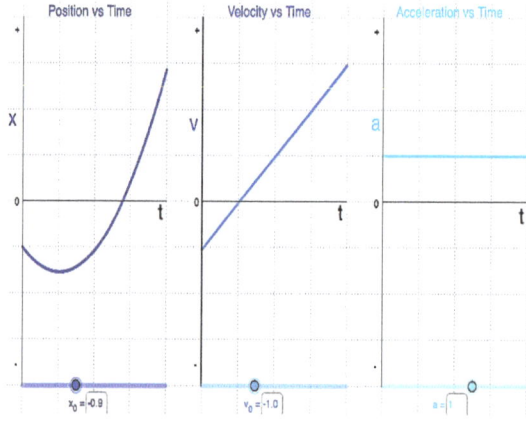

To better appreciate the relation between parabolic distance, linear velocity and constant acceleration, try the GeoGebra activity "Impact of Constant Acceleration".

2.3. INTEGRALS

Strange Staircases

Linear velocity $v = at + b$ is neat. It can capture any constant rate of change. Distance traveled equals the signed area of rectangles and triangles lying between v and the t-axis. Let us recall how we deduce it:

- We raise or lower an otherwise constant velocity in steps that touch true velocity from below but never cross it.
- The area of the associated staircase, measured from the horizontal axis, sets a lower bound for distance traveled.
- We create a second set of steps that touch true velocity from above.
- The area of its staircases sets an upper bound for distance traveled.
- By making the steps narrower and narrower, upper and lower bounds converge to the answer.

This method works for any continuous function $f(x)$. Although the staircases can look strange, they do what we want. The area of an upper bound staircase (upper sum) never understates the net displacement, while area of a lower bound staircase (lower sum) never overstates it.

As steps get narrower, upper and lower sums squeeze toward the same answer: the net signed area between $f(x)$ and the x-axis. The staircases are known as Riemann slices. Their limit is called the integral.

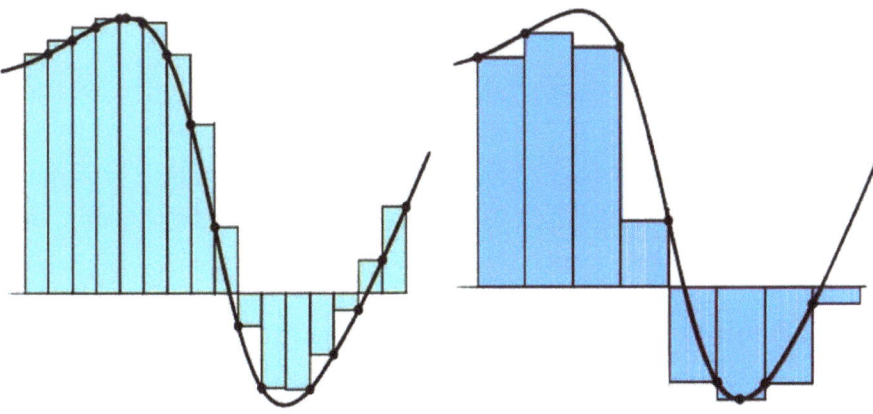

CHAPTER 2

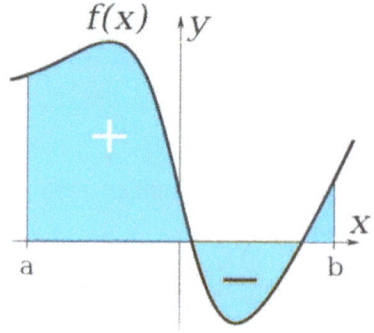

For maximum clarity, we need to specify the x region we're building staircases on. When we do, the integral is said to be definite. The picture to the left illustrates the definite integral of $f(x)$ from $x = a$ to $x = b$.

We already did this kind of calculation for sledding. When the velocity is $0.8t$, the sled travels $0.4t^2$ from time 0 to time t. To say this using calculus terminology, the distance traveled $0.4t^2$ is the definite integral of velocity $0.8t$ from 0 to t.

Integrals as Functions

Ordinary functions f map each input number x to an output number y. In contrast, definite integrals map each input function f to an output function F. To calculate $F(b)$ for even a single b, we need to know everything f does from b back to the initial a. Moreover, each a defines a different F. How can we express this more clearly?

To express a long sequence of variables in a tractable way, pick a single letter to describe each string and tack on integers to distinguish within the string: say, $x1, x2, x3, \ldots$. To emphasize that the integers are part of the variable name, attach them as subscripts: $x_1, x_2, x_3, \ldots$.

There are two main ways to express the sum of a long sequence. The first is to write $x_1 + x_2 + \cdots + x_n$, where the dots mean "fill in the rest of this pattern". The second is to use the large Greek capital letter Σ ("sigma") as a shorthand for summation, so that $\sum_{k=1}^{n} x_k$ denotes the sum of x_k for all integer k between 1 and n.

To apply this to staircases, mark $n+1$ points from $x_0 = a$ to $x_n = b$ to bound n steps. Each step S_k spans x_{k-1} to x_k with width $\Delta x_k = x_k - x_{k-1}$.

Denoting by $f_k^{\max}$ and $f_k^{\min}$ the maximum and minimum $f(x)$ values for x in S_k, the upper and lower sums are $\sum_{k=1}^{n} f_k^{\max} \Delta x_k$ and $\sum_{k=1}^{n} f_k^{\min} \Delta x_k$ respectively. The limit as step widths approach zero is the definite integral from a to b.

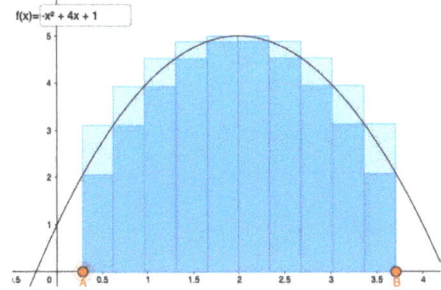

The GeoGebra activity "Riemann Sums" displays upper and lower Riemann sums and their definite integral limits. Users choose the function, the range of integration, and the number of sub-intervals.

Integral Notation

Leibniz notation $\int_a^b f(x)\,dx$ is standard for definite integrals. The $f(x)\,dx$ are infinitesimal counterparts to the various $f(x_k)\Delta x_k$. The subscripts get dropped because there are too many to describe. A stretched old-German S replaces Σ. Leibniz notation is fine unless we take it literally. A definite integral is not an actual summation but a limit that refers to the area that stretches vertically from the x–axis to $f(x)$ and horizontally from $x = a$ to $x = b$.

Interestingly, $\int_a^b f(x)\,dx$ does not require $a < b$. If the x_k get smaller rather than larger, each nudge Δx_k will be negative rather than positive. The net effect switches the sign compared to an interval running from b to a, In other words,

$$\int_a^b f(x)\,dx = -\int_b^a f(x)\,dx \text{ for all } a, b, \text{ and } f.$$

To use the definite integral as a function, pick some lower bound a and define $F(x) = \int_a^x f(z)\,dz$, where z ranges from a to x. Often this is written using the same letter for z and x, which is sloppy but does no damage, as there is no other reasonable way to interpret the integral.

CHAPTER 2

While each lower bound a generates a different definite integral F, all values of $F(x)$ shift by the same amount. Specifically, a shift to lower bound A increases F by

$$\int_A^x f(z)\,dz - \int_a^x f(z)\,dz = \int_A^a f(z)\,dz,$$

which does not depend on x. Our earlier game of fetch provided an example. To convert the distance from thrower to the distance from initial position of the ball, just subtract the distance from thrower to ball. All other area calculations stay the same.

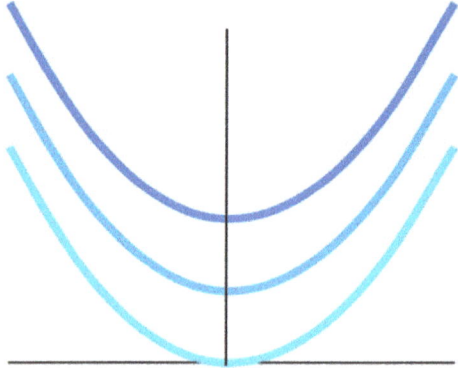

Graphically, adding a constant to F makes a parallel shift. Since parallel shifts don't change shapes, we can view them as keeping functions in the same family.

An indefinite integral, written $\int f(x)\,dx$ or $\int^x f(z)\,dz$, refers to this family. It picks a particular F and adds an unspecified constant C. For any such F, the definite integral from a to b equals $F(b) - F(a)$ since the C cancels out. Hence, converting between definite and indefinite integrals is typically easy. When you're not sure which type is needed, it's safer to include C until it cancels out or is adjusted to achieve a required value for the integral.

Jumps

Not all the functions we want to integrate are continuous throughout. Recall puppy fetching a ball. In our unrealistic model, velocity jumps at time 10 from $+4$ to -4. How does the jump affect integration?

For definite integrals, it doesn't matter. Just add the relevant signed areas between each velocity segment and the time axis. The jump in velocity shows up as a kink in the chart of puppy going and coming.

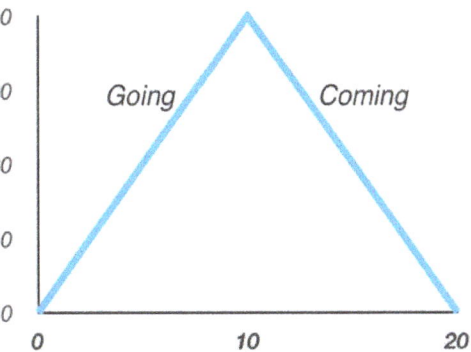

Indefinite integrals are trickier. Looking at each segment alone, the going velocity has indefinite integral $4t + C_{going}$ while the coming velocity has indefinite integral $-4t + C_{coming}$. No Riemann integral can jump, as no nudge of width 0 can add nonzero area. Hence we need the two integrals to match at $t = 10$, which requires C_{coming} to equal $C_{going} + 40$.

This approach works just as well on functions with multiple jumps, known as piecewise continuous. For definite integrals, just add the signed areas "under" each continuous piece. For indefinite integrals, either add a common C to the definite integrals or calculate indefinite integrals for each piece and adjust the constants to make the integrals match at boundaries.

Alternatively, we can approximate the jumps by steep inclines, as if the dotted lines were slightly slanted. For our purposes, continuous functions are sufficiently versatile that we won't bother much with jumps.

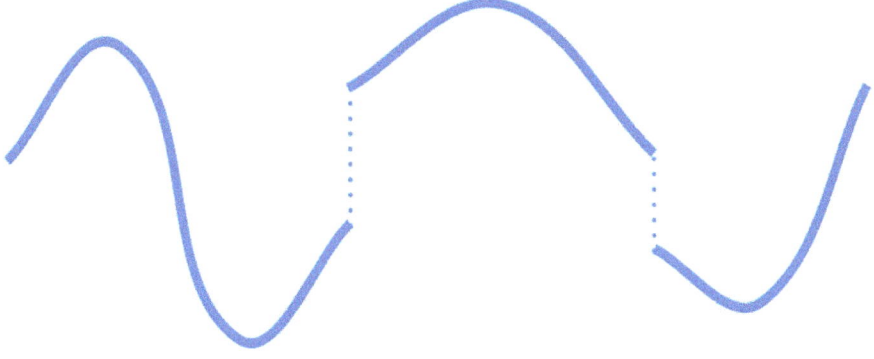

2.4. FUNDAMENTAL THEOREM

Nudges of Sums

What happens to a sum of nudges when we nudge it? It grows by a nudge. No big deal, we might think. Wrong. The growth nudge is the key link between integrals and derivatives.

The darker area below depicts a definite integral $F(x)$ for some lower boundary a, while the lighter area depicts the nudge $F(x+\Delta x) - F(x)$. The lighter area is at least its width Δx times its minimum height $f(x+\Delta x)$ and at most Δx times its maximum height $f(x)$.

The edges don't always determine the maximum and minimum heights. Nevertheless, provided f is continuous at x, the two bounds squeeze toward $f(x)\Delta x$ as Δx gets small. More precisely, the average height squeezes toward $f(x)$, which we can express formally as

$$\lim_{\Delta x \to 0} \frac{F(x+\Delta x) - F(x)}{\Delta x} = f(x).$$

Does the expression on the left-hand side look familiar to you? It should. $F(x)$ and $F(x+\Delta x)$ are function names for y and $y+\Delta y$ respectively. That makes the numerator Δy, and $\lim_{\Delta x \to 0} \frac{\Delta y}{\Delta x}$ is just the derivative defined earlier. We can write this as

$$\frac{dF(x)}{dx} = \frac{d}{dx}\int f(x)\,dx = f(x).$$

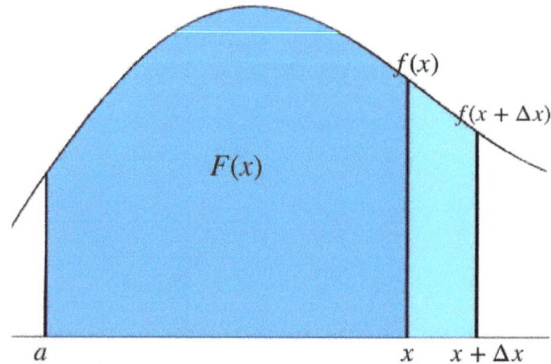

In words, any continuous function equals the derivative of its integral. This key identity is known as the first Fundamental Theorem of Calculus.

SUMS OF NUDGES

The Theorem indicates that differentiation reverses integration. This is similar to how subtraction reverses addition or division reverses multiplication. However, calculus is more sensitive to reversal order than arithmetic is. When we differentiate a function and integrate it back, the result can differ by a constant C from the original:

$$\int \frac{dF(x)}{dx} dx = F(x) + C.$$

This explains why an indefinite integral is also called an antiderivative. In words, the derivative of F integrates back to F, give or take a constant. This is known as the second Fundamental Theorem of Calculus.

Cubes and Pyramids

Charted below is the area under a parabola that starts off flat at the origin and passes through $(x, 3x^2)$. To its right is an ice cube with sides x and volume $V = x^3$. How are the two related, apart from color? Let's apply the Fundamental Theorem to find out.

Imagine that the ice cube is lying in the corner of a freezer and that water vapor is condensing evenly on its three open faces, with extension Δx on each side. Since each face starts with area x^2, the total nudge ΔV in volume is $3x^2 \Delta x$ plus nudges of nudges $3x(\Delta x)^2$ along three edges and $(\Delta x)^3$ on the corner. Taking the limit, V has derivative $3x^2$. Applying the second Fundamental Theorem, the area under the parabola from the origin to x equals V. Hence

$$\frac{dx^3}{dx} = 3x^2 \quad \text{and} \quad \int_0^x 3z^2 dz = x^3.$$

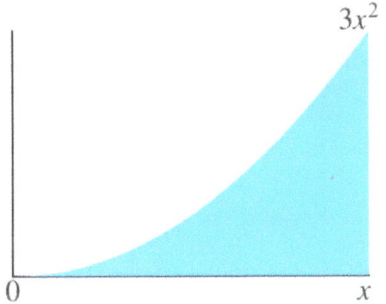
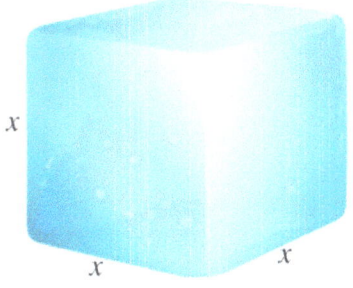

55

CHAPTER 2

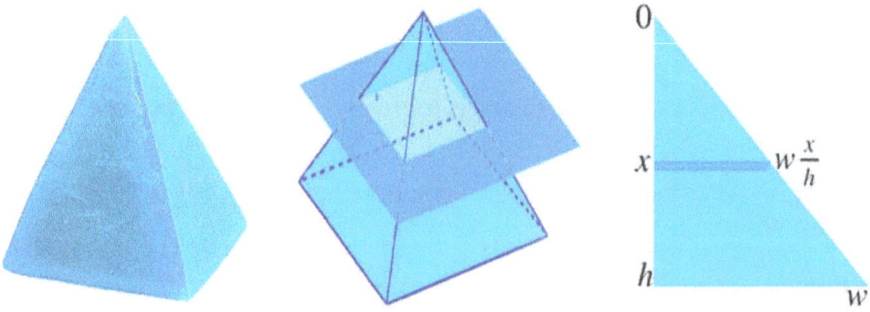

For another application, consider a square pyramid with base width w and height h. As we move up the pyramid, sides taper off linearly to zero. A cross section x units from the top will have sides $(w/h)x$ and area $(w/h)^2 x^2$. Charted against x, area matches the parabola x^2 apart from the scaling factor $(w/h)^2$. It follows that the antiderivative is $F(x) = \frac{1}{3}(w/h)^2 x^3 + C$ and the volume is $F(h) - F(0) = \frac{1}{3} w^2 h$. In words, volume equals one-third of base area times height.

Here is another way to obtain the same result. Fasten every vertex of a hollow cube to the vertex farthest away. The diagonals will cross at center and outline six identical square pyramids, as displayed below on the left. The picture to its right colors the pyramids and separates them. Given base width w, each pyramid is $\frac{1}{2}w$ high with volume c Take one of the pyramids, slice it into square cross sections Δw thick, and then expand or compress each thickness by a factor of $2h/w$. This changes overall height to h, changes the volume to $\frac{1}{3} w^2 h$, and in the limit of tiny Δw preserves the pyramid shape.

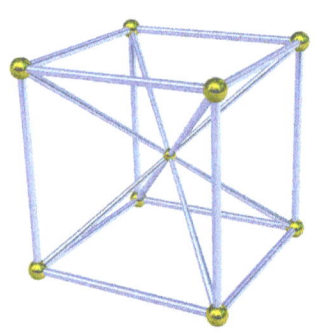
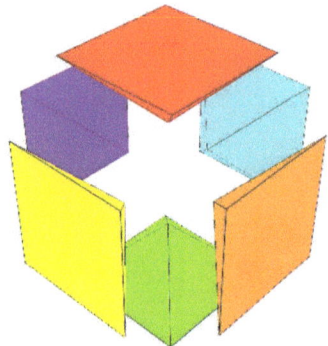

We can analyze any pyramid this way. Since each side scales linearly with height, cross sectional area A scales quadratically. Hence the volume, as integral of A, always equals one-third of base area times height.

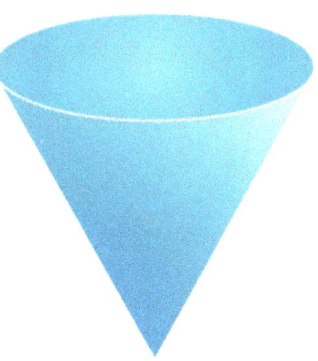

All the shape changes is the formula for base area. For example, a cone cup h high with base radius r has base area πr^2 and volume $\frac{1}{3}\pi r^2 h$. A rectangular pyramid with base width w and length l has base area wl and volume $\frac{1}{3}wlh$.

More generally, suppose two three-dimensional objects have cross sections at height h of different shape but matching area. If that is true for all h, the objects will have the same volume. This is known as Cavalieri's principle and we will see some applications shortly.

Spheres and Cones

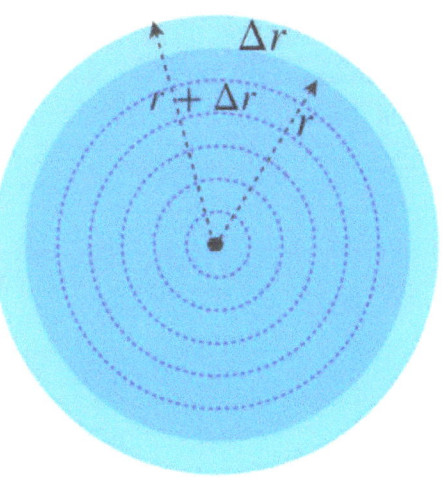

Suppose mathnesia strikes us again over circles. This time we recall the area formula $A(r) = \pi r^2$ given radius r but not the circumference formula $B(r)$.

Luckily, seeing an onion slice reminds us that circles can be split into thin concentric rings. A ring with inner radius r and outer radius $r + \Delta r$ has area of approximately $B(r)\Delta r$. Applying the first Fundamental Theorem,

$$B(r) = \frac{dA(r)}{dr} = 2\pi r.$$

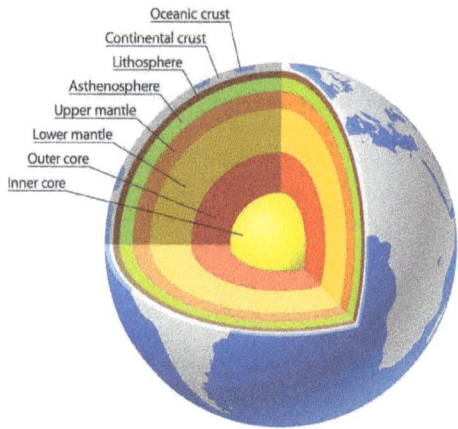

Similarly, we can think of spheres as nested layers of shells. The nest will resemble this model of Earth's core, except that each layer should be extremely thin. A shell of radius x with surface area $S(x)$ and thickness Δx has volume of close to $S(x)\Delta x$. Hence total volume $V(r)$ must equal the integral:

$$V(r) = \int_0^r S(x)dx.$$

If we know that $S(r) = 4\pi r^2$, we can readily calculate $V(r) = \tfrac{4}{3}\pi r^3$. Alternatively, if we know $V(r)$, we can differentiate it to obtain $S(r)$. Calculus saves us from memorizing both surface area and volume, as each formula generates the other.

Now comes the hard part. How do we know that either formula is correct? The neatest solution was discovered nearly 2400 years ago. It equates a sphere's volume to a cylinder's volume less a cone's volume when all three objects have the same height and maximal diameter. To prove it, replace the cone with two half-height cones of the same total volume, stack them tip to tip, and slide them into the cylinder. Then compare the areas of horizontal slices at the same height, like in the charts below.

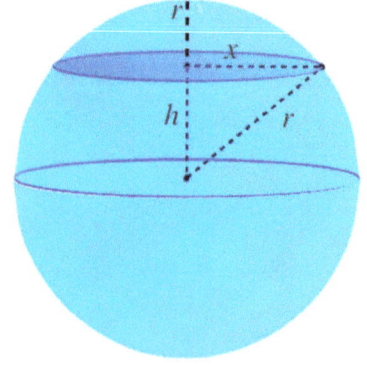
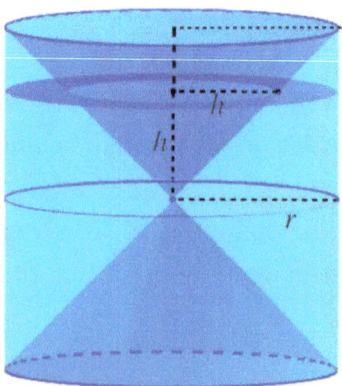

At height h above or below the center, the cylinder slice has area πr^2 and the cone slice has area πh^2, so the ring outside the cone has area $\pi r^2 - \pi h^2$. By the Pythagorean theorem this matches the area πx^2 of the corresponding slice of the sphere. Since this is true for every h, the sphere volume equals the difference $\frac{4}{3}\pi r^3$ of the cylinder volume $2\pi r^3$ and the volume $\frac{2}{3}\pi r^3$ of the two cones.

Using nudge tricks to avoid formal calculus can strengthen our intuition. However, one merit of calculus is that it can work even when intuition partly fails. For example, suppose that the chart of a double cone inside a cylinder confuses us, but we understand that slicing a sphere at height h exposes an internal surface area of $\pi(r^2 - h^2) = \pi r^2 - \pi h^2$. Integrate both terms from $h = -r$ to $h = r$: the first gives the volume of the cylinder while the second gives the volume of a double cone. The analysis is the same as before even if we miss the geometric connections.

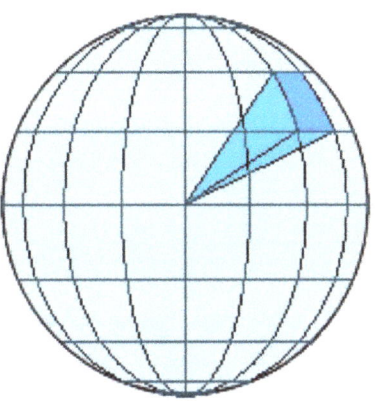

Here is another way to relate surface area to volume. Carve a sphere into tiny near-pyramids P_i, each with base at surface and tip at center. Since each P_i has a volume of $\frac{1}{3}r$ times its base area, sphere volume is approximately $\frac{1}{3}r$ times total base area $S(r)$. In the limit, $S(r) = (3/r)V(r) = 4\pi r^2$.

Wedding Bands

Instead of squeezing a sphere into a cylinder, imagine we drill a cylindrical hole of radius R through the center of a sphere of radius $r > R$. The remaining portion, known as a spherical ring, has height $2h = 2\sqrt{r^2 - R^2}$. It resembles a napkin ring when h is close to R and a wedding band when h is much smaller than R. Holding h fixed, how does the ring's volume change as r expands?

CHAPTER 2

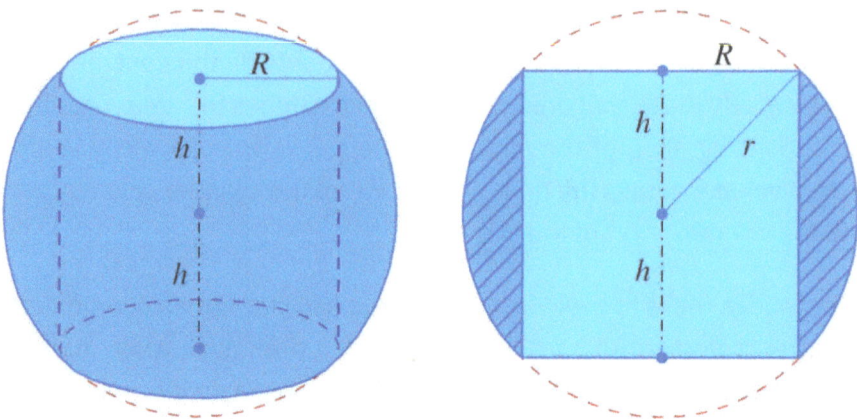

At first glance, the volume must increase as the ring gets wider. However, the outside gets flatter so the ring gets thinner. Surprisingly, these two effects exactly cancel out: only h matters, not R. How do we know? Consider a ring slice at height x from the center. Since its outer radius is $\sqrt{r^2 - x^2}$ and its inner radius is R, its area is

$$\pi(r^2 - x^2) - \pi R^2 = \pi(r^2 - R^2 - x^2) = \pi(h^2 - x^2).$$

This doesn't depend on r for any x. Hence the volume must equal $\frac{4}{3}\pi h^3$, as if the band were a sphere of radius h with $R = 0$.

For example, a spherical ring of solid gold that is one centimeter high and one meter in diameter sounds far more massive than it is. Its volume of half a cubic centimeter, comparable to that a small marble, will contain not quite 10 grams of gold, worth about $\$750$ at current market prices. However, it would be much thinner than a human hair, with maximum thickness of about 25 microns (millionths of a meter).

Is that important to remember? Of course not. But being able to imagine small nudges and adjust them is important. The modern world depends on manufacturing to much higher precision. Gold leaf less than 0.2 microns is used widely and some astronauts' space helmets were coated with gold leaf so thin at 0.05 microns that it was partially transparent.

3

Slopes

Slopes are to derivatives what areas are to integrals. Finding points where slopes are zero helps to identify optimal solutions. Four core rules simplify calculation tremendously. A short formula covers every power of x.

3.1. TANGENT LINES

Ant on a Hill

While building a fence for puppy, I noticed an ant carting a blade of grass bigger than itself. At least I think it was an ant; the blade mostly obscured it. That got me wondering how much we can learn from blades of grass about who's holding them up and where they're heading. While I didn't make much progress on the who, an answer leapt out on the where: slope. The slope of the blade helps signal the direction of motion.

To appreciate why, imagine a narrow path only one ant wide that advances in the .x.-direction up a quadratic hill $y = x(7 - x)$. If we treat the ant as infinitesimal, the blade passes though the current point (x_0, y_0) and lies flush against the path. As there is only one way the blade can lie flush—it can't wobble around without changing its point of contact—we can determine exactly where it's pointing.

CHAPTER 3

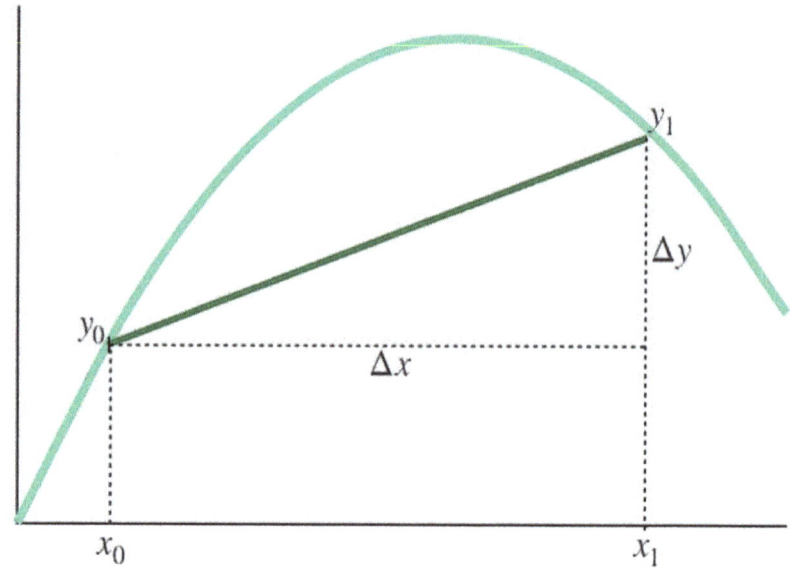

How do we know the blade can't wobble? Imagine we drill a straight borehole from (x_0, y_0) to some other point (x_1, y_1) on the specified path. The borehole's slope equals the ratio $\Delta y/\Delta x$ of the rise $\Delta y = y_1 - y_0$ to the run $\Delta x = x_1 - x_0$. Subtract $y_0 = x_0(7 - x_0)$ from $y_1 = x_1(7 - x_1)$ to obtain

$$\Delta y = 7(x_0 + \Delta x) - (x_0 + \Delta x)^2 - 7x_0 + x_0^2 = 7\Delta x - 2x_0\Delta x - (\Delta x)^2,$$

which implies $\Delta y/\Delta x = 7 - 2x_0 - \Delta x$. Since Δx is nonzero, the borehole's slope can never equal $7 - 2x_0$ but can get infinitesimally close. The only slope left for a line flush to the hill is exactly $7 - 2x_0$, which matches the derivative at that point.

A line segment connecting two points on a graph is called a secant, while a line segment flush to a graph is called a tangent. For any $y_0 = f(x_0)$ and $y_1 = f(x_1)$, the slope of the secant is $\dfrac{y_1 - y_0}{x_1 - x_0}$. As x_1 approaches x_0, the limit if it exists indicates the slope of the tangent at (x_0, y_0) and is written

$$\left.\frac{dy}{dx}\right|_{x=x_0} \quad \text{or} \quad \left.\frac{df(x)}{dx}\right|_{x_0}.$$

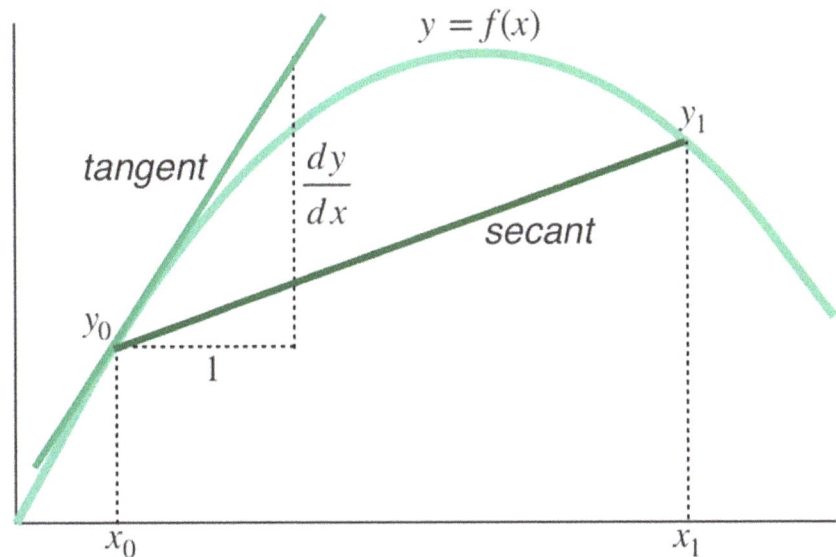

How can we say this more crisply? Newton placed dots over functions to denote derivatives, which makes $\dot{f}$ the derivative of f. Clearly $\dot{f}$ is a function too. Laplace notation, which is more common, shifts the dot right and converts it to a slash called prime. We can then summarize our core finding as:

The tangent to f at x_0 has slope $f'(x_0)$, if the latter exists.

Informally, the first Fundamental Theorem of Calculus says that the area under f has slope f, while the second Fundamental Theorem says that the area under the slope of f equals f plus a constant.

Again, don't let the names and symbols rattle you. Play with the GeoGebra activity "Slope Grapher" to build intuition for the concepts.

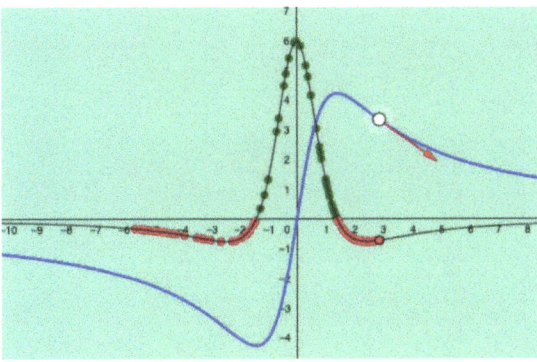

CHAPTER 3

Existence

As we learned in Chapter 2.3, every continuous or piecewise continuous function generates definite integrals throughout its domain. Derivatives are less tolerant. They can't stand jumps or kinks.

To see what this means for tangent lines, let's draw some. Below left is a step function from Chapter 2.2. Before the jump, the slope is clearly zero, and the tangent matches the left part. After the jump, the slope is also zero, and the tangent matches the right part. At the jump, the tangent has zero slope for calculations that stay on the same side and infinite slope for calculations that span both sides.

Below right is the integral of the step function. While continuous, it is kinked at top. Before the kink, the slope is positive and the tangent matches the left side. After the kink, the slope is negative and the tangent matches the right side. At the kink, the tangent can have any slope in between.

Hence there is no tangent line at jumps and there are multiple tangent lines at kinks. Making the left and right segments curved rather than straight doesn't change these conclusions. Fortunately, most functions we will meet are smooth enough to have derivatives everywhere. The rest have few jumps or kinks in the ranges we care about.

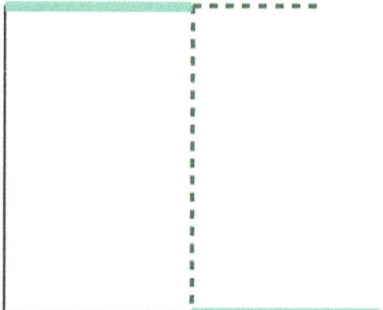
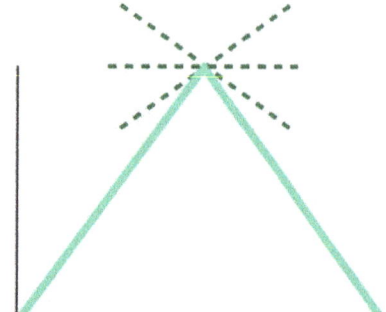

Maxima and Minima

The top of a function hill must have a horizontal tangent, because no nearby points can be higher. Hence any derivative at the top must be zero. For example, the quadratic hill $y = x(7-x)$ reaches its maximum value at $x^* = 3.5$, where the derivative $7 - 2x^*$ equals zero. This was the answer to the rectangular fence problem in Chapter 1.1.

Similarly, the bottom of a function hill always has a horizontal tangent line, because no nearby points can be lower. Any derivative at the bottom, like at the top, must be zero to match the slope of the tangent. At first glance, the rectangular fence violates this condition. To minimize area, the fence needs to be 0 wide or 0 long. The corresponding derivatives seem to be or -7. What went wrong?

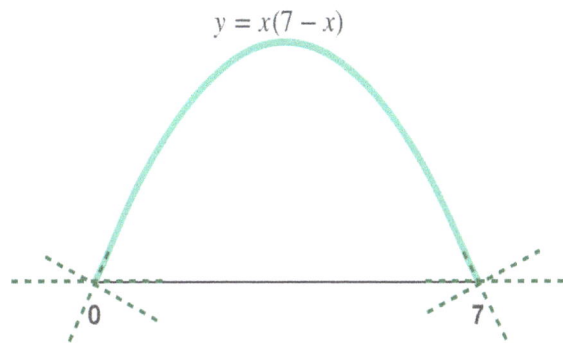

The problem is the boundary conditions imposed by nature. Since no side of the fence can be negative, x must lie between 0 and 7. At either boundary, one tangent line gives way to many.

In general, only the interior maxima or minima of smooth functions (that is, without jumps or kinks) need to equate derivatives to zero. Boundaries have tangent possibilities similar to kinks.

These insights yield an incredibly useful rule:

If a problem can be formulated as choosing x to maximize or minimize a smooth $f(x)$, any interior solution x^* must satisfy $f'(x^*) = 0$.

CHAPTER 3

For an unusual application, imagine an ant carrying a blade of grass seeks the highest point on a camel. Measuring the height of every point on the camel will take close to forever. It is better to use the blade-must-be-horizontal rule to sift down to a handful of candidates.

Any point with derivative 0 is known as stationary, since the function there isn't trending up or down. Stationary points are usually either local maxima, where nothing infinitesimally close is higher, or local minima, where nothing infinitesimally close is lower.

Given several local maxima or minima, how can calculus identify the overall global maximum or minimum? It can't. Any derivative is a local measure only. At best calculus can tell us—sometimes—that there is only one stationary point and if so, whether it marks a global maximum or global minimum. Later we'll learn how.

Some functions have multiple stationary points but no global maximum or minimum. Here is a clipped graph of $y = x^3 - 3x$, whose slope is $3x^2 - 3$. The point $(-1, 2)$ is a local maximum while $(1, -2)$ is a local minimum. Yet any $x > 2$ will generate $y > 2$, while any $x < -2$ will generate $y < -2$.

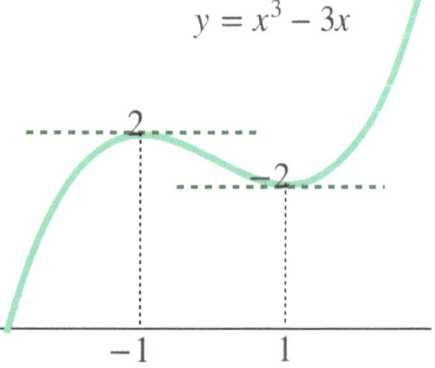

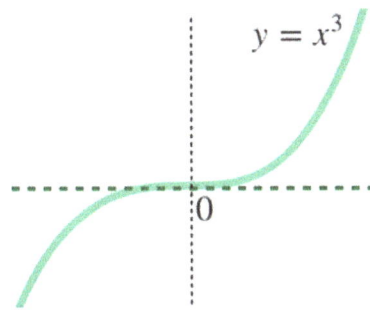

Some stationary points are neither local maxima nor local minima. The graph of $y = x^3$ provides an example. The derivative is $3x^2$, so the curve flattens out at the origin. The horizontal tangent lies above everything to the left of the axis and below everything to the right.

However, if $f(a) = f(b)$, a continuous f must have at least one minimum or maximum between a and b. This is known as Rolle's Theorem. Intuitively, since the path ends at the height it started at, whatever goes up must come down and vice-versa. By extension, any continuous f that crosses the x-axis multiple times must have at least one local extreme between each crossing.

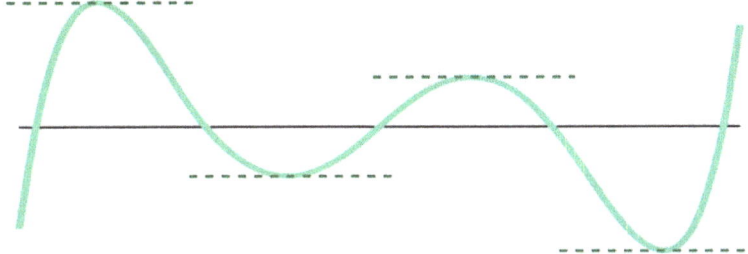

Exercise your calculus vocabulary in this crossword puzzle.

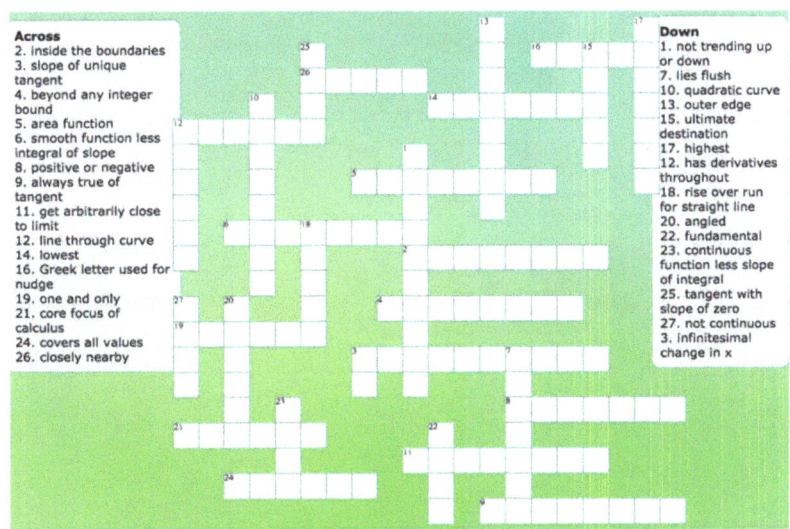

Across
2. inside the boundaries
3. slope of unique tangent
4. beyond any integer bound
5. area function
6. smooth function less integral of slope
8. positive or negative
9. always true of tangent
11. get arbitrarily close to limit
12. line through curve
14. lowest
16. Greek letter used for nudge
19. one and only
21. core focus of calculus
24. covers all values
26. closely nearby

Down
1. not trending up or down
7. lies flush
10. quadratic curve
13. outer edge
15. ultimate destination
17. highest
12. has derivatives throughout
18. rise over run for straight line
20. angled
22. fundamental
23. continuous function less slope of integral
25. tangent with slope of zero
27. not continuous
3. infinitesimal change in x

CHAPTER 3

3.2. REFLECTORS

Tennis Solitaire

Have you ever played tennis against a wall? I have and it's annoying. My worst shots punish me by rebounding even more off course. To remedy, I'm drawing up plans for a game I'll call tennis solitaire. It's played inside a big cylinder. If the player in the center hits a straight shot in any direction, the ball should bounce back toward the center.

Do you agree? Let's think about it mathematically. Although the wall is curved, the tennis ball will find any spot it hits nearly flat, with slope equal to the tangent. For any ball hit horizontally, denote by A and B the two angles it makes with the tangent. If the cylinder were flipped over, A and B would reverse, yet the circular cross section would look the same. Hence A and B must be equal right angles and a tennis ball hit straight from the center should bounce straight back.

Given a point (x, y) on the wall, the radial line to the origin has slope y/x, which equals the run-to-rise ratio for angle C. The tangent's slope is the negative of the rise-to-run ratio for angle D. Since the tangent is perpendicular to the radial line, D must equal C. Hence the tangent has slope $-x/y$.

To convert this into terms involving one variable only, recall that $x^2 + y^2 = r^2$ for a circle of radius r. Hence $y = \sqrt{r^2 - x^2}$ and we can write

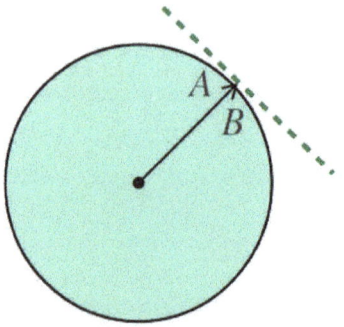
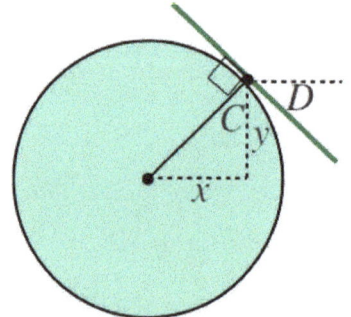

the derivative as $\left(\sqrt{r^2 - x^2}\right)' = -x/\sqrt{r^2 - x^2}$. Congratulations—we just solved a not-so-simple calculus problem without using calculus.

Over 2000 years ago, Archimedes used the geometry of circle tangents to calculate the surface area of a sphere. Enclose a hollow sphere in a hollow cylinder of the same diameter $2r$ and height $2r$. Cut a horizontal slice Δh wide through both to form two narrow ribbons. Archimedes deduced that the ratio of ribbon areas converges to 1 as Δh shrinks.

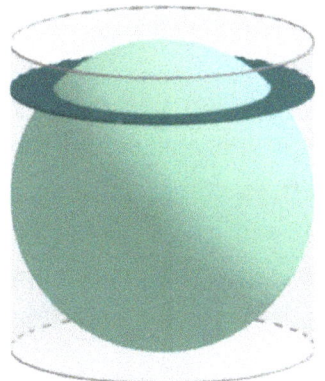
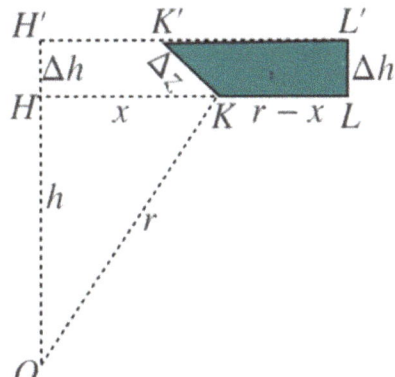

Let's confirm this. The dark thin ring on the left extends from sphere to cylinder. The ribbon that passes through $L'L$ has circumference $2\pi r$ and width Δh, while the ribbon that passes through $K'K$ has circumference $2\pi x$ and width close to Δz. Since $K'K$ is tangent to the sphere, it is perpendicular to radial line KO. Hence angles HOK and HKK' match. Equating the ratios of height to hypotenuse,

$$\frac{\Delta h}{\Delta z} = \frac{x}{r} \quad \text{or} \quad \frac{2\pi r \cdot \Delta h}{2\pi x \cdot \Delta z} = 1.$$

While the sphere bends slightly away from its tangent, the gap gets minuscule as Δz approaches zero. Since this is true at any height h, the sphere must have the same surface area $4\pi r^2$ as the hollow cylinder. While Chapter 2.4 proved this result in two other ways, Archimedes' proof is more direct and intuitively appealing.

CHAPTER 3

Ellipses

Perfect symmetry under rotation gives circles swelled heads. Their boasts of superiority annoy other 2D figures and trigger fights at the geometric saloon. After a fight, a unit circle with equation $X^2 + Y^2 = 1$ may stagger out with new coordinates $x = aX$ and $y = bY$ for some positive a and b. Unless $a = b$, the circle is squashed, with maximum width $2a$ and maximum height $2b$. The squashed circle is called an ellipse, with equation

$$\frac{x^2}{a^2} + \frac{y^2}{b^2} = 1.$$

For every x or y that satisfies this equation, so do $-x$ or $-y$, which implies symmetry about both x-axis and y-axis. As Chapter 7 will explain, rotation maps x to $(x - uy)/\sqrt{1 + u^2}$ and y to $(y + ux)/\sqrt{1 + u^2}$, where u denotes the new slope of what was the horizontal axis. This generalizes the equation to $\frac{(x - uy)^2}{a^2} + \frac{(y + ux)^2}{b^2} = 1 + u^2$. To simplify analysis, let us rotate ellipses back to $u = 0$. Let us also switch x with y as needed to keep $a \geq b$, so that the ellipse is no higher than it is wide.

Elliptical walls are perfect for a game I'll call tennis tandem. Have player 1 stand at a special point F_1 called a "focus" that is $c = \sqrt{a^2 - b^2}$ to the right of center. Have player 2 stand c to the left at focus F_2. No matter where either player hits the ball, absent crazy spin, it should bounce to the other player. Moreover, the ball always travels distance $2a$ between them, making for a nice steady game.

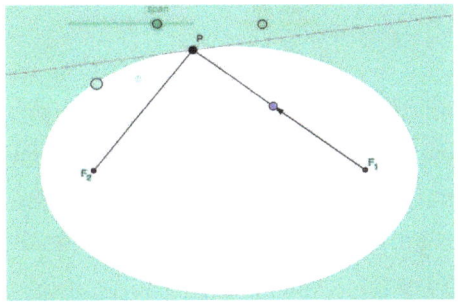

The GeoGebra activity "Elliptical Reflection" shows how this works.

To prove the constant distance property, let $D_1 = \sqrt{(x-c)^2 + y^2}$ denote the distance from focus $F_1 = (c, 0)$ to a point $P = (x, y)$ on the ellipse. Substitute $y^2 = b^2 - (b/a)^2 x^2$ to obtain

$$D_1 = \sqrt{\frac{a^2 - b^2}{a^2} x^2 - 2cx + c^2 + b^2} = \sqrt{\frac{c^2}{a^2} x^2 - 2cx + a^2} = a - \frac{cx}{a},$$

which is positive since $c < a$ and $|x| \le a$. For the distance D_2 from P to $F_2 = (-c, 0)$, the calculation is the same except for the sign on c. Hence $D_1 + D_2 = 2a$ as claimed.

Next let's imagine the tangent line at P is a mirror. For R the reflection of F_2, suppose there is a point Q on the tangent that makes F_1QR shorter than F_1PR. In that case F_1QF_2, which is just as long as F_1QR, will be shorter than $D_1 + D_2 = 2a$, so that Q lies inside the ellipse instead of on the tangent. Hence F_1PR must be straight, in which case angle B equals angle A as required for reflection.

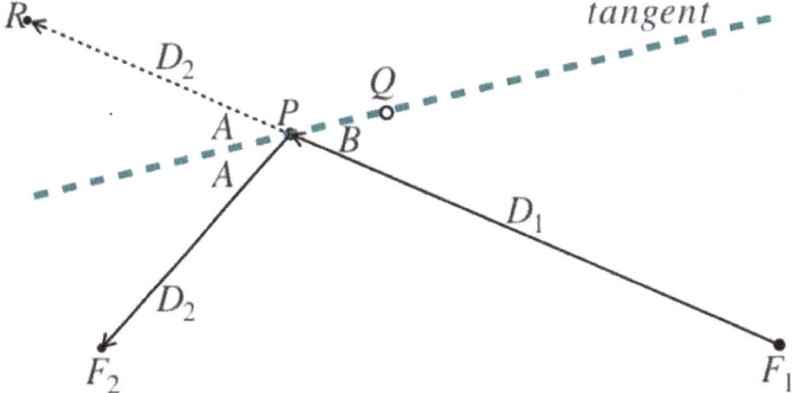

Parabolas

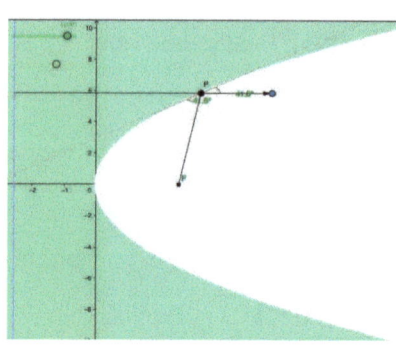

Imagine an ellipse that is extremely long. Any tennis balls hit from one focus will bounce nearly parallel to the long axis. In the limit the ellipse never closes at the far end and turns into a parabola. Explore its behavior in the GeoGebra activity "Parabolic Reflection".

An upward-facing parabola based at the origin has equation $y = cx^2$ with $c > 0$ and slope $2cx$. How high is the focus F? At point G where projection switches from horizontal to vertical, the tangent must be angled at $45°$ for a slope of 1. This requires $x = 1/(2c)$, which implies height $F = 1/(4c)$.

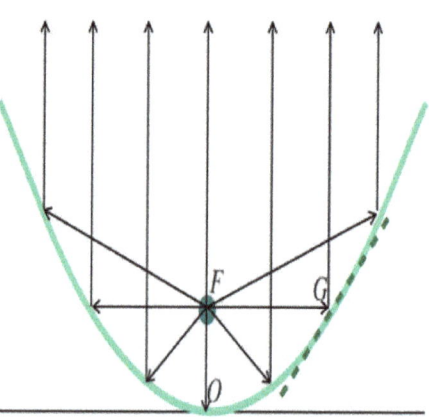

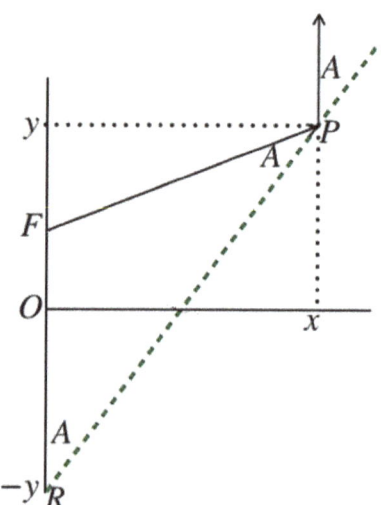

How about other projections from F? The tangent to any point $P = (x, y)$ on the parabola must pass through $R = (0, -y)$ to have slope $2cx$. It is easily checked that FP and FR have equal lengths. This makes RFP an isosceles triangle, with angle at R that matches the angle A of incidence and reflection. Hence all projections from F reflect straight up.

Parabolic reflectors are extraordinarily useful. They help flashlights and headlights project light forward. They help telescopes concentrate weak light from outer space. They help microwave receivers relay information around the globe. What a testament to the power of tangents!

Hyperbolas

Suppose we flip the sign in the basic ellipse equation to form a difference of squares $x^2/a^2 - y^2/b^2 = 1$. The resulting graph is called a hyperbola. Like an ellipse, it is symmetric about both axes. However, it splits into two branches since $|x|$ cannot be less than a. Since $|y|/b$ must be less than $|x|/a$, the hyperbola is also bounded by lines $y = (b/a)x$ and $y = -(b/a)x$. These lines are asymptotes, which means that the hyperbola gets infinitesimally close to them as $|x| \to \infty$.

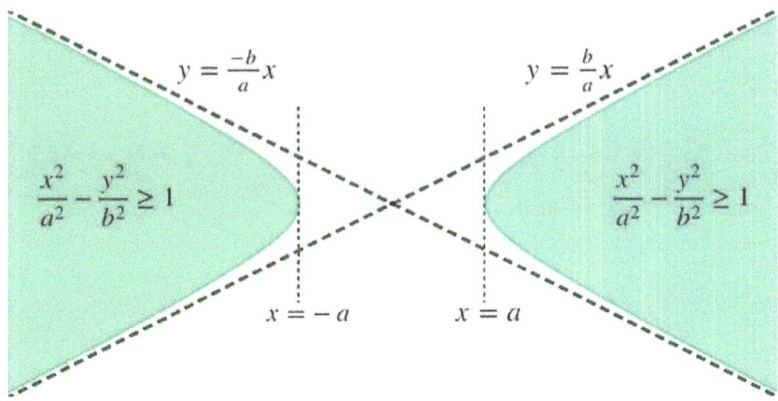

Rotation generalizes the defining equation to $\frac{(x-uy)^2}{a^2} - \frac{(ux+y)^2}{b^2} = 1 + u^2$. When $a^2 = b^2 = 2$ and $u = -1$ this simplifies to $xy = 1$, with x- and y-axes as asymptotes.

Like ellipses, hyperbolas have two foci F_1 and F_2, with distances D_1 and D_2 to them that are closely related. However, D_1 and D_2 differ by $2a$ for hyperbolas, whereas they sum to $2a$ for ellipses. Also, the distance $2c$ between F_1 and F_2 equals $2\sqrt{a^2 + b^2}$ instead of $2\sqrt{a^2 - b^2}$. Apart from signs, the algebraic proof matches that for ellipses.

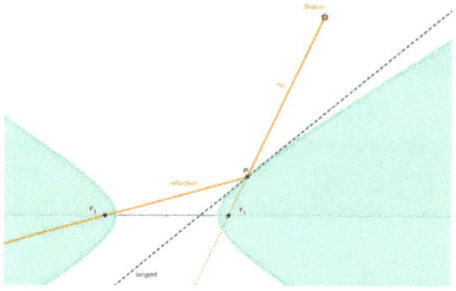

Hyperbolas possess a peculiar reflective property. As the $\overline{GF_2}$ GeoGebra activity "Hyperbolic Reflection" demonstrates, they redirect all light beams heading for one focus toward the other.

For geometric proof, pick any point P on the hyperbola. Choose G on $\overline{F_1P}$ such that $GP = PF_2$. For midpoint M between G and F_2, the line through $\overline{PM}$ cuts the extensions of $\overline{F_1P}$ and $\overline{F_2P}$ into four slices with equal angles at P. A ray heading to F_2 will reflect to F_1 if and only if $\overline{PM}$ is tangent to the hyperbola.

To prove tangency, we next show that any point Q on the extended $\overline{PM}$ lies outside the hyperbola. By the triangle inequality, $F_1G + QG > QF_1$. $QG = QF_2$ because QM is a perpendicular bisector of . $F_1G = PF_1 - PF_2 = 2a$ because P lies on the hyperbola. Combining these relations indicates $QF_1 - QF_2 < 2a$, confirming the claim.

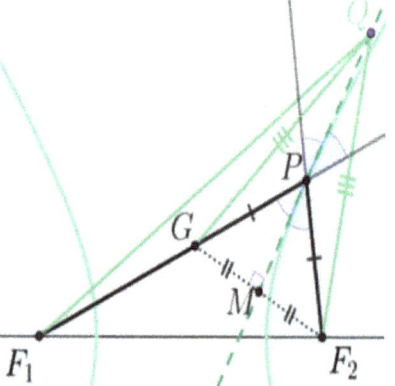

Conic Sections

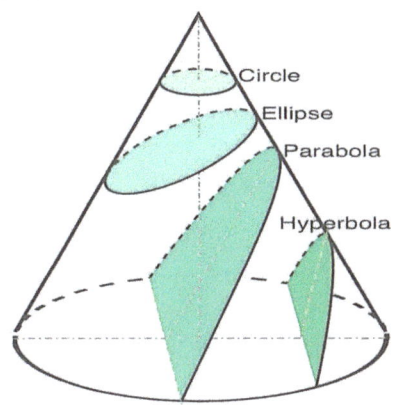

Every circle, ellipse, parabola or hyperbola can be viewed as a cross section of a double cone stacked tip to tip. The slope of the cross section relative to the cone determines the shape: circle if parallel to the base, ellipse if shallower than the side of the cone, parabola if just as steep as the side, and hyperbola if steeper.

To demonstrate this, let's start with a right-angled double cone defined by $x^2 + y^2 = z^2$ and a two-dimensional plane defined by $z = sx + q$ for every y. They intersect where $(1-s^2)x^2 - 2sq\,x + y^2 = q^2$, which clearly describes a circle for $s = 0$ and a parabola for $s = 1$. For other s, define $w = 1/(1-s^2)$ and rewrite the intersection as

$$\frac{(x-qws)^2}{q^2w^2} + \frac{y^2}{q^2w} = 1.$$

For $0 < s < 1$ this is a standard ellipse shifted to a center $(qws, 0)$ with $a = |qw|$ and $b = |q|\sqrt{|w|}$. For $s > 1$ this is a standard hyperbola shifted to the same center with the same a and b.

A general conic section satisfies $(x-h)^2 + (y-k)^2 = (sx + py + q)^2$, which rearranges into the form

$$Ax^2 + Bxy + Cy^2 + Dx + Ey + F = 0.$$

Given a feasible F, every such equation generates a hyperbola when $B^2 > 4AC$, parabola when $B^2 = 4AC \ne 0$, ellipse when $B^2 < 4AC$ and circle when $B = A - C = 0$. This is easily shown when $B = 0$. When $B \ne 0$, rotation with slope $u = \dfrac{A-C}{B} + \sqrt{\left(\dfrac{A-C}{B}\right)^2 + 1}$ will eliminate the xy term while leaving the value of $B^2 - 4AC$ unchanged.

3.3. FOUR CORE RULES

Unit and Addition Rules

In arithmetic, the concepts are simple but computation is tedious and we need to memorize addition and multiplication tables to do it efficiently. The concepts of basic calculus are harder to grasp but computation is relatively easy. All we need to remember are four core rules, which come in both integral and derivative forms.

The first two core rules we've already used. The unit rule is obvious once we understand the notation.

Unit Rule: $\int_0^x 1\,dz = x$ and $x' = \dfrac{dx}{dx} = 1$.

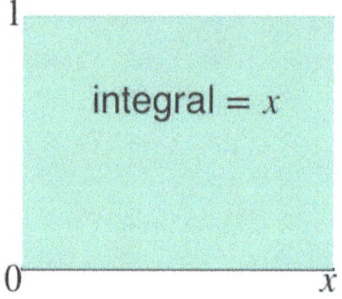
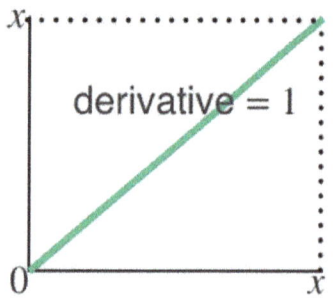

The first form of the unit rule equates the integral of 1 from 0 to x with the area of a 1-by-x block. The second form says the derivative is 1 when something changes one-for-one with itself. These boil down to the identity rules of multiplication and division.

The addition rule says that the integral of a sum equals the sum of the integrals and the derivative of a sum equals the sum of the derivatives:

Addition Rule: $\int_a^b (f(x) + g(x))\,dx = \int_a^b f(x)\,dx + \int_a^b g(x)\,dx$

and $(f+g)' = \dfrac{d}{dx}(f+g) = \dfrac{df}{dx} + \dfrac{dg}{dx} = f' + g'$.

Both forms follow from the distributive rule $(f+g)\Delta x = f\Delta x + g\Delta x$. The integral is the limit of sums while the derivative is the limit of ratios.

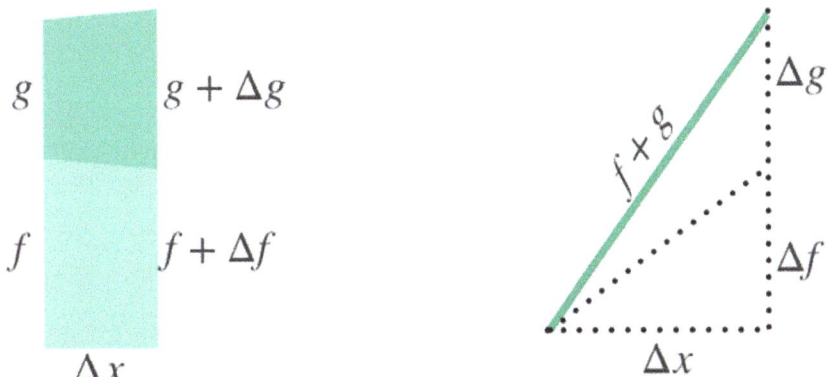

For a simple but important consequence, set $g=0$, which means that $g(x)=0$ for all x. Since $f = f+0$, $0'=0$ and $\int_a^b 0\,dx = 0$ for all a and b. It follows that no f has two distinct derivatives f' or two antiderivatives that differ by a non-constant. Why not? Because $f=g$ implies $f-g=0$, with derivative 0 and some constant antiderivative C.

Chain Rule

A composite function plugs outputs of one function machine $y = g(x)$ into another function machine $z = f(y)$. It is written $f(g(x))$ or $f \circ g$. For example, if $g(x) = x+1$ and $f(y) = y^2$, $f(g(x)) = (x+1)^2$. The chain rule provides shortcuts for differentiating composite functions. It also facilitates changes of variables inside integrals. In the expressions below, note that $f'(g(x))$ means f' evaluated at $g(x)$.

Chain Rule: $\int_{g(a)}^{g(b)} f(y)\,dy = \int_a^b f(g(x)) g'(x)\,dx$

and $(f \circ g)' = \dfrac{d}{dx} f(g(x)) = f'(g(x)) \cdot g'(x) = f'g'$.

Alternatively, we can write $\int_{g(a)}^{g(b)} f\,dg = \int_a^b f \dfrac{dg}{dx}\,dx$ or $\dfrac{d}{dx}(f \circ g) = \dfrac{df}{dg} \cdot \dfrac{dg}{dx}$. To prove the derivative version, define $\Delta y = g(x+\Delta x) - g(x)$ and $\Delta z = f(y + \Delta y) - f(y)$, and calculate

$$\frac{d}{dx} f(y) = \lim_{\Delta x \to 0} \frac{\Delta z}{\Delta x} = \lim_{\Delta x \to 0}\left(\frac{\Delta z}{\Delta y} \cdot \frac{\Delta y}{dx}\right) = \lim_{\Delta y \to 0} \frac{\Delta z}{\Delta y} \cdot \lim_{\Delta x \to 0} \frac{\Delta y}{\Delta x} = f'(y) \cdot g'(x).$$

CHAPTER 3

For example, when $y = x+3$ and $f(y) = y^2$, the chain rule tells us

$$\frac{dy^2}{dx} = \frac{dy^2}{dy} \cdot \frac{dy}{dx} = 2y \cdot 1 = 2(x+3).$$

To check, expand $y^2 = (x+3)^2$ to $x^2 + 6x + 9$ and differentiate using the addition rule to obtain $2x + 6 = 2(x+3)$.

For a more interesting example, consider the square root function $y = \sqrt{x}$, which by convention is never negative. For any positive x, $x = y^2$. Plug that into the chain rule to obtain

$$1 = \frac{dx}{dx} = \frac{dy^2}{dy} \cdot \frac{dy}{dx} = 2y \cdot \frac{dy}{dx} = 2\sqrt{x} \cdot \frac{d\sqrt{x}}{dx},$$

which implies $\left(\sqrt{x}\right)' = 0.5/\sqrt{x}$. Now we've learned something new!

To generalize, let f^{-1} denote the function that maps $y = f(x)$ back to x. It is called the inverse of f. The derivatives of f and f^{-1}, if they exist, are reciprocals. Why? The chain rule implies

$$1 = x' = \left(f^{-1} \circ f\right)' = \frac{df^{-1}(y)}{dy} \cdot \frac{df(x)}{dx}.$$

Hence $\frac{df^{-1}(y)}{dy} = \frac{1}{f'(x)}$, or more simply $\frac{dx}{dy} = 1 / \frac{dy}{dx}$.

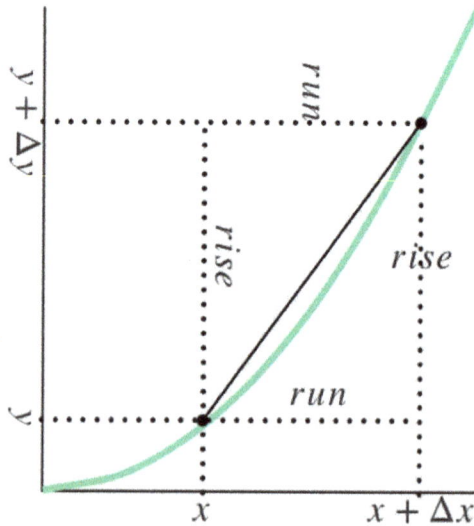

This result has a simple geometric interpretation. The graph of x versus y is just the graph of y versus x with axes switched. Hence their slopes are reciprocals: the run-to-rise $\frac{\Delta x}{\Delta y}$ versus the rise-to-run $\frac{\Delta y}{\Delta x}$.

The unit rule, addition rule, and chain rule allow us to derive the slopes of conic sections without invoking geometry. For a circle $x^2 + y^2 = a^2$, differentiate both sides with respect to x to see that

$$0 = \frac{da^2}{dx} = \frac{dx^2}{dx} + \frac{dy^2}{dy} \cdot \frac{dy}{dx} = 2x + 2y\frac{dy}{dx}, \text{ which implies } \frac{dy}{dx} = \frac{-x}{y}.$$

The same approach applied to the ellipse $x^2/a^2 + y^2/b^2 = 1$ indicates that $\frac{dy}{dx} = \frac{-b^2 x}{a^2 y} = \frac{-bx}{a\sqrt{a^2 - x^2}}$. The corresponding hyperbola slope is $\frac{bx}{a\sqrt{a^2 + x^2}}$.

Substitute $x = (a/b)z$ to transform the ellipse into the circle $z^2 + y^2 = b^2$. For $y > 0$ the ellipse area is $\int_{-a}^{a} 2y\,dx$ while the circle area is $\int_{-b}^{b} 2y\,dz$. The chain rule shows that the former is a/b times the latter:

$$\int_{-a}^{a} 2y\,dx = \int_{-b}^{b} 2y \frac{dx}{dz}\,dz = \frac{a}{b} \int_{-b}^{b} 2y\,dz = \frac{a}{b}(\pi b^2) = \pi ab.$$

For a derivation of sphere volume using the chain rule, imagine a sphere of radius r centered at the origin is divided into nested layers of hollow vertical cylinders. A cylinder of radius z will touch the sphere at heights $h = \sqrt{r^2 - z^2}$ and $-h$. Given its circumference $2\pi z$, its contribution to volume is $4\pi hz$ times its infinitesimal thickness. Hence the total volume is $\int_{0}^{r} 4\pi hz\,dz$. Since h and z jointly describe a circle, $\frac{dz}{dh} = \frac{-h}{z}$. If we substitute $\frac{dz}{dh}\,dh$ for dz and change the limits to $h(0) = r$ and $h(r) = 0$, the integral is transformed into $\int_{r}^{0} 4\pi(-h^2)\,dh = -4\pi \cdot \tfrac{1}{3} h^3 \big|_{r}^{0} = \tfrac{4}{3} \pi r^3$ .w

Product Rule

The product rule for calculus is the hardest for beginners to grasp, because it doesn't work like the arithmetic version. Let's start with the derivative form by itself:

Product Rule: $(fg)' = \frac{d}{dx}(fg) = f(x)\frac{dg}{dx} + g(x)\frac{df}{dx} = fg' + gf'$.

CHAPTER 3

To prove this, define $\Delta f = f(x+\Delta x) - f(x)$ and $\Delta g = g(x+\Delta x) - g(x)$. Dropping x from the notation, $(f + \Delta f)(g + \Delta g)$ sums the base product fg, two nudges $f\Delta g$ and $g\Delta f$, and a nudge of nudge $\Delta f \Delta g$. Hence

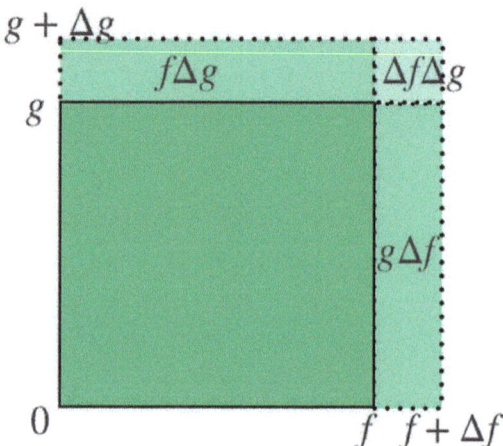

$$(fg)' = f \cdot \lim_{\Delta x \to 0} \frac{\Delta g}{\Delta x} + \lim_{\Delta x \to 0}(g + \Delta g) \cdot \lim_{\Delta x \to 0} \frac{\Delta f}{\Delta x}$$

$$= fg' + (g+0)f' = fg' + gf'.$$

In words, the derivative of a product equals the sum of two products: the first function times the derivative of the second and the second function times the derivative of the first.

Like the other core rules, the product rule has many useful implications, which are often treated as rules of their own. For example, when $f(x)$ is a constant c,

$$(cg)' = cg' + gc' = cg' + g \cdot 0 = cg',$$

which includes $(cx)' = c$ as a special case. For another simple application, setting $f(x) = g(x) = x$ indicates that

$$(x^2)' = xx' + xx' = 2x.$$

Now for something more complex. Define $h(x) = 1/g(x)$ for some smooth g that does not pass through zero. Since $0 = 1' = (hg)' = hg' + gh'$, $h' = -hg'/g = -g'/g^2$. Applying the product rule to fh,

$$\frac{d}{dx}\left(\frac{f}{g}\right) = fh' + hf' = \frac{-f}{g^2}g' + \frac{1}{g}f' = \frac{gf' - fg'}{g^2}.$$

This is known as the division rule. While many students try to memorize this separately from the product rule, I hope you won't. It is too easy to mix things up. When in doubt, take a bit longer and apply the product rule multiple times.

The integral form of the product rule says that $\int (fg)'dx$ equals $\int fg'dx + \int gf'dx$. If we rearrange and apply the chain rule, we obtain a procedure known as integration by parts. The expression below uses the notation $h\big|_a^b = h(b) - h(a)$ and $\int_{h(a)}^{h(b)} dh = \int_a^b h'(x)dx$.

Integration by Parts: $\int_{g(a)}^{g(b)} f\, dg = fg\big|_a^b - \int_{f(a)}^{f(b)} g\, df$.

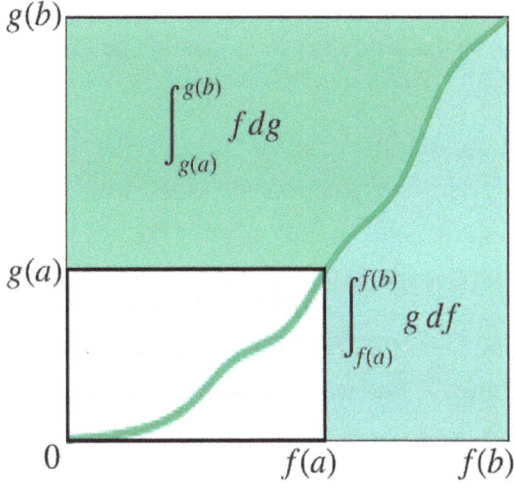

To visualize integration by parts, graph the points $(f(x), g(x))$. $\int_{g(a)}^{g(b)} f\, dg$ is the area between the graph and the g-axis. $\int_{f(a)}^{f(b)} g\, df$ is the the area between the graph and the f-axis. Their sum equals $fg\big|_a^b$, the area of the big rectangle less the area of the small rectangle.

For example, when $f(x) = x^2$ and $dg = dx$, $\int_0^b x^2 dx = x^3\big|_0^b - \int_0^b 2x^2 dx$, which confirms that $\int_0^b x^2 dx = \tfrac{1}{3}x^3\big|_0^b = \tfrac{1}{3}b^3$. However, integration by parts is rarely that straightforward. There are many ways to carve a given integral into f and dg parts and the $\int g\, df$ rearrangements they imply are often messier than the original. It takes a lot of practice to recognize good ways to carve.

3.4. POWERS OF X

Induction

Athelia the Hunny is my favorite barbarian princess. She could have been the Huns' most glorious leader. Unfortunately, she lost out to her brother Attila on a misunderstanding.

"What are your global ambitions?" the Council of the Huns asked its would-be chiefs. *"What hungers you most as Hun?"*

"I won't rest until we conquer the Roman Empire!" said Attila, and the crowd cheered.

"I just want the land we control and the land next to it," said Athelia, and the crowd fell silent. The Huns were stunned.

Poor Athelia. They didn't appreciate her daring. Should the Huns conquer neighbors, the land they control would expand. Yet it would still border land they didn't control. Athelia would want that too. She wouldn't stop until she conquered the whole continent or died trying.

Thankfully, Athelia's spirit lives on in math, under the name of induction. Suppose a claim refers to a number n, and we know that

P1: The claim is true when $n=1$.

P2: Whenever the claim is true for n, it is true for $n+1$.

Then induction says the claim holds for every positive integer. To prove this, let M be the smallest positive integer for which the claim fails. Thanks to **P1**, M is at least 2 and the claim holds for $M-1$. Thanks to **P2**, the claim holds for $(M-1)+1=M$ after all.

Now that we know what induction is, let's apply it to verify that

$$(x^n)' = nx^{n-1} \text{ for all positive integers } n.$$

Let's start by making sure the formula itself is clear. Since x^n means $x \cdot x \cdots x$ for n occurrences of x, $x^k x^m = x^{k+m}$ and $(x^k)^m = x^{km}$ for any positive integers k and m. Defining $x^0 = 1$ extends that to cover $n = 0$. Hence, the claim is $x' = 1$, $(x^2)' = 2x$ and so on.

Clearly the claim starts out true, confirming **P1**. To prove **P2**, let's assume the claim is true for a given n. Applying the product rule to $f(x) = x$ and $g(x) = x^n$ show that the claim is true for $n+1$:

$$\frac{dx^{n+1}}{dx} = x\frac{dx^n}{dx} + x^n\frac{dx}{dx} = x \cdot nx^{n-1} + x^n = nx^n + x^n = (n+1)x^n.$$

Negative Integer Powers

Let's extend the handy rule $x^m x^n = x^{m+n}$ to cover powers that are negative integers. Since $x^{-n} x^n = x^0 = 1$, x^{-n} must equal $1/x^n$. Calculating

$$\frac{x^m}{x^n} = \frac{x^{m-n} x^n}{x^n} = x^{m-n} = x^m x^{-n}$$

confirms that multiplying by x^{-n} does everything that dividing by x^n does.

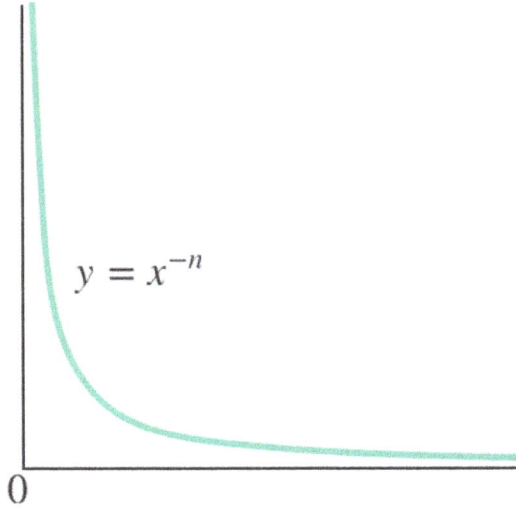

$y = x^{-n}$

For $n > 0$, x^n starts at the origin, grows to 1 at $x = 1$, and keeps growing without bound as x grows. Hence x^{-n} is infinite at $x = 0$, shrinks to 1 at $x = 1$, and keeps shrinking toward 0. This makes the $x-$ and $y-$ axes asymptotes for x^{-n}.

For n a positive integer, $y = x^{-n}$ has a second curve left of the origin: a mirror image $(-x)^{-n} = x^{-n}$ for n even and a $180°$ rotation $(-x)^{-n} = -x^{-n}$ for n odd. There is always a gap at $x = 0$.

The product rule implies $0 = 1' = (x^n x^{-n})' = x^n (x^{-n})' + x^{-n}(x^n)'$, while direct calculation indicates that $(x^n x^{-n})' = 1' = 0$. Rearranging terms,

$$(x^{-n})' = -x^{-2n}(x^n)' = -nx^{n-1-2n} = -nx^{-n-1}.$$

If we define $m = -n$, this simplifies to the same form as for positive integers. It also applies for $m = 0$. Hence

$$\frac{dx^n}{dx} = nx^{n-1} \text{ for any integer } n \text{ and } x \neq 0.$$

Rational Powers

The mth root of x is inverse to the mth power. That is, it equals a number y for which $y^m = x$. It can be written $\sqrt[m]{x}$, although the square root usually drops the identifier $m = 2$. If m is even and x is negative, $\sqrt[m]{x}$ has no real number solution. If m is even and x is positive, there are negative and positive solutions with the same absolute value. Whenever there is potential confusion, we will presume x and $\sqrt[m]{x}$ are positive.

An alternative expression for $\sqrt[m]{x}$ is $x^{1/m}$, which extends the rule $(x^k)^m = x^{km}$ to cover any pair of rational numbers. In particular, $(x^{1/m})^m = (x^m)^{1/m} = 1$, which makes the inverse relation very clear.

Like other pairs of inverse functions, the graphs of $x^{1/m}$ and x^m switch the $x-$ and $y-$ axes without making any other changes. While all positive roots and powers start at $(0,0)$ and pass through $(1,1)$, higher m accentuates the curve: the roots stay closer to 1 while the powers surge more near 1.

The inverse pairing makes the slope of $y = x^{1/m}$ reciprocal to the slope of $x = y^m$. Hence for any integer m,

$$\frac{dy}{dx} = \frac{dy}{dy^m} = \frac{1}{my^{m-1}} = \frac{1}{m}y^{1-m} = \frac{1}{m}(x^{1/m})^{1-m} = \frac{1}{m}x^{1/m-1}.$$

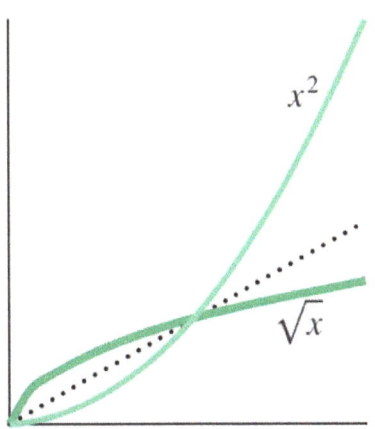
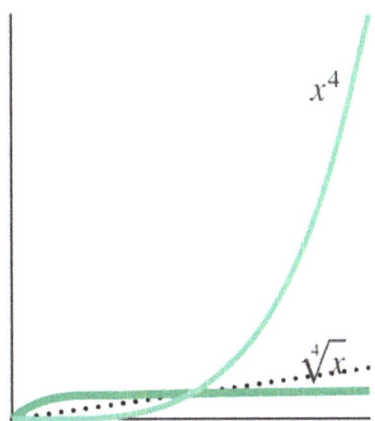

How about the derivative of $y^n = x^{n/m}$, where n is an integer too? Applying the chain rule,

$$\frac{dy^n}{dx} = \frac{dy^n}{dy} \cdot \frac{dy}{dx} = ny^{n-1} \cdot \frac{1}{m} x^{1/m-1} = \frac{n}{m} x^{n/m-1/m+1/m-1} = \frac{n}{m} x^{n/m-1}.$$

Since any rational number equals the ratio n/m of two integers,

$$\frac{dx^p}{dx} = px^{p-1} \text{ for any rational number } p.$$

Irrational Powers

Any number on the number line can be expressed as a decimal expansion, like $\sqrt{2} = 1.4142135\ldots$. The dots at the end mean "and so on" and suggest "fill in the rest of the pattern yourself". Only we can't, as there is no recognizable pattern.

Rational numbers have decimal expansions that end in infinitely repeating sequences, like 000… 333…, or 157157…. Irrational numbers never do. How do we know $\sqrt{2}$ is irrational? If rational, there must be two integers m and n with ratio $m/n = \sqrt{2}$, and we can keep dividing them by .2. until at least one is odd. Since $m^2 = 2n^2$, m can't be odd. That makes m^2 divisible by 4, which makes n^2 divisible by 2, so n can't be odd either.

Any number r can be viewed as the limit of a sequence of rational numbers. For example, the decimal expansions of $\sqrt{2}$ converge to the true $\sqrt{2}$ as they get longer. Intuitively, x^r should equal the limit of a sequence of rational powers, which suggests that

$$\frac{dx^r}{dx} = rx^{r-1} \text{ for any real number } r.$$

That turns out to be correct. We will defer the proof until Chapter 6, where we rewrite x^r in terms of exponentials and logarithms.

Polynomials

Polynomials are sums of constant multiples of whole-number power functions. The two most important types are linear $ax+b$ and quadratic ax^2+bx+c. Polynomials are cubic if the highest power is three and quartic if the highest power is four. Eventually we tire of special names and call a polynomial n^{th}-order if the highest power is n.

Calculating derivatives and integrals of polynomials is super-easy. Wherever we see x^m, replace it with mx^{m-1} for a derivative or $x^{m+1}/(m+1)$ for an integral, and then simplify the coefficients. For example, $2x^3 + 4x$ has derivative $6x^2 + 4$ and integral $½x^4 + 2x^2 + C$.

It is also easy to calculate derivatives of the ratio of two polynomials or any power of the ratio. Just apply the chain rule. In practice this often gets messy. Fortunately, online resources like Wolfram Alpha can quickly compute any derivative.

Integrals of ratios of polynomials, let alone fractional powers of ratios, are more complicated. Most of them don't have solutions that can be expressed as ratios of polynomials. Even there, however, polynomials can provide useful approximations. We'll learn more about them later.

4

Slopes of Slopes

The second derivative is the slope of the slope. Its sign distinguishes between maxima and minima. Acceleration is the second derivative of displacement and relates directly to force applied. Good design often relies on smooth changes to second derivatives.

4.1. CURVINESS

Playpen with Cover

The playpen for puppy was such a success that I want to build one for kitten. This one will need a cover to keep kitten from climbing out. I'll stick four poles at the corners of an x-by-w rectangle, extend them h high, and wrap see-through mesh around the sides and top. How can I maximize the interior volume $V = xwh$ given the mesh area A?

For a really clever solution, lay an imaginary mirror on the ground and observe that the playpen and its reflection form a completely covered box. To maximize total box volume given surface area, our fence-making logic from Chapter 1 tells us to form a cube. The cube's dimensions will be $x = w = 2h$, with total volume x^3 and total surface area $6x^2$. The playpen's share will be exactly half, so $A = 3x^2$ or $x = \sqrt{A/3}$.

Since I'm not that clever, let's rethink the problem from scratch. The mesh must cover two sides measuring x by h, two sides of w by h, and a top of x by w. Hence $A = 2(x+w)h + xw$. Given any h and xw, V is fixed but A is minimized by setting $w = x$, in which case $A = 4xh + x^2$. This reduces our problem to the maximization of $V = x^2 h = \frac{1}{4}(Ax - x^3)$, which is stationary at $A = 3x^2$.

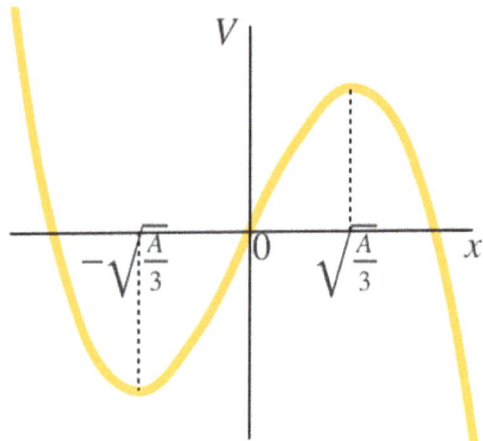

To check, let us graph V. Oops, something is wrong. There's no global minimum or maximum and two stationary points. Aha, x must be positive. Moreover, the graph provides a hint. It curves up around the first stationary point and curves down around the second. What does that tell us?

For a hint about the hint, let's watch our imaginary ant friends. When they climb over smooth hills, the grass stalks on their backs point up initially, level out at top, and point down after. When they climb through smooth valleys, the stalks point down initially, level out at bottom, and point up after.

Translated into mathematical terms, this means that slopes decrease around maxima and increase around minima. The slope of the slope helps distinguish the two cases. If negative, the stationary point is a local maximum. If the slope of the slope is positive, the stationary point is a local minimum.

In this case the slope of V is $\frac{1}{4}(A - 3x^2)$. The slope of the slope is $-\frac{3}{2}x$, which is negative for $x > 0$ and positive for $x < 0$. Hence $x = \sqrt{A/3}$ sets a local maximum while $x = -\sqrt{A/3}$ sets a local minimum.

Second Derivative

The derivative of the derivative is called the second derivative, which makes the ordinary derivative the first derivative. Leibniz notation simplifies expressions like $\frac{d}{dx}\left(\frac{df}{dx}\right)$ to $\frac{d^2f}{dx^2}$. Caution: don't confuse the second derivatives with a first derivative with respect to x^2!

Using Newton's dot notation, $\ddot{f}$ is the second derivative of f. The corresponding Laplace notation is f''. Using Laplace notation and calling a nudge $-\Delta x$ instead of the usual Δx, we can define

$$f''(x) = \lim_{\Delta x \to 0} \frac{f'(x) - f'(x - \Delta x)}{\Delta x}.$$

Replacing the f' terms with their limiting ratios shows that

$$f''(x) = \lim_{\Delta x \to 0} \frac{f(x + \Delta x) - 2f(x) + f(x - \Delta x)}{(\Delta x)^2}.$$

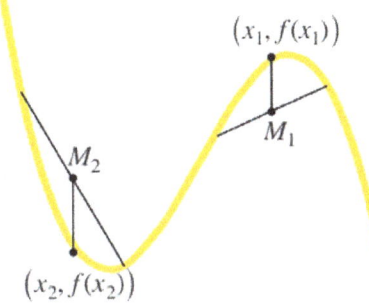

Here is a geometric interpretation. A secant from $(x - \Delta x, f(x - \Delta x))$ to $(x + \Delta x, f(x + \Delta x))$ has a midpoint M of $(x, \tfrac{1}{2}f(x - \Delta x) + \tfrac{1}{2}f(x + \Delta x))$. Hence the numerator is twice the signed distance from $(x, f(x))$ to M.

Convexity

If $(x, f(x))$ lies above all nearby secants, $f''(x)$ cannot be positive. If it lies below all nearby secants, $f''(x)$ cannot be negative. Hence $f'(x) = 0$ marks a local minimum if $f''(x) > 0$ and a local maximum if $f''(x) < 0$.

Like for f', no particular value for f'' can identify global extremes. However, a f'' that never changes sign reveals a lot about the global properties of f:

CHAPTER 4

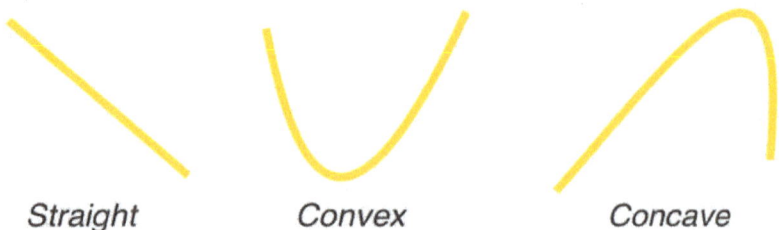

- If $f'' = 0$ everywhere, f is a straight line and has a stationary point only if it is constant.
- If $f'' > 0$ everywhere, f bends upward and is called convex. Since all secants lie above it, any stationary point must be the unique global minimum.
- If $f'' < 0$ everywhere, f bends down ward and is called concave. Since all secants lie below it, any stationary point must be the unique global maximum.

We can generalize convexity and concavity to allow for kinks and straight bits. Kinks don't weaken the properties of being above or below the secants but straight segments do. Flats at top or bottom generate a continuous range of global maxima or minima.

Inflection Points

Points where $f''(x) = 0$ are known as inflections. They tell us where curves have straightened out. Usually the curve bends upward on one side of the inflection and downward on the other. Sometimes the inflection is just a lull before the previous bend resumes.

A stationary inflection point with $f''(x) = f'(x) = 0$ can be a maximum, a minimum, or neither. For example, x^3, $|x|^3$ and $-|x|^3$ are the same curves apart from signs on one side or the other of the origin. Each is stationary

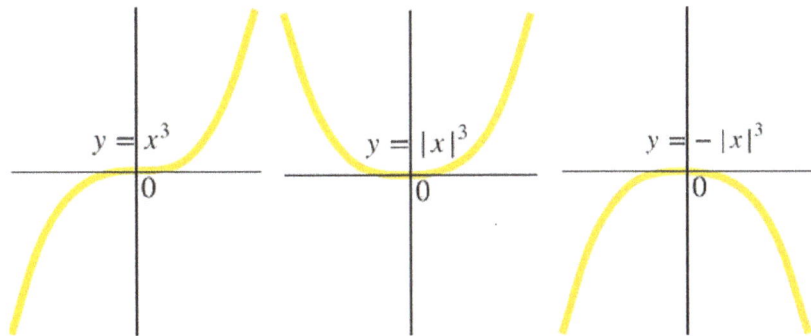

and inflected at the origin. Yet the origin is a global minimum for $|x|^3$, a global maximum for $-|x|^3$, and neither for x^3. Only the derivatives nearby can give us the information we need.

Double Integrals

When we know f'' and want to figure out f, we need to integrate twice. This can be written $f(x) = \int^x \left(\int^z f''(w)dw \right) dz$ or $f(x) = \int \int^x f''(z) dz dx$. The simplest case is $f'' = 0$, which indicates that f is straight but doesn't tell us its slope C_1 or intercept C_2. Here the first antiderivative is C_1 and the second antiderivative is $\int^x C_1 dz = C_1 x + C_2$.

Even when f'' isn't zero, f includes a $C_1 x + C_2$ term that can't be specified without more information. For example, in the case of a sled with position $f(t)$ at time t, $f''(t)$ tells us only how the speed $f'(t)$ is changing over time. To calculate $f'(t)$, we need to know the initial speed $f'(0)$. We then add the initial position $f(0)$ to determine $f(t)$.

Degrees of Smoothness

If f consists of continuous pieces, there is no limit to how many times we can integrate it in succession. Applying the Fundamental Theorem, we can then differentiate the integral the same number of times to regenerate the original f. However, this doesn't guarantee that f can be differentiated even once throughout an interval. The most we can say is that any derivative f' must be continuous.

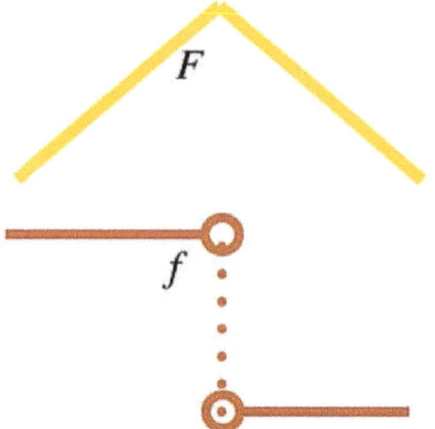

For example, the integral F to the left satisfies $F' = f$ yet no $F'' = f'$ exists at the kink. At best we generate the step function below it that isn't defined at the jump. Does that matter? In most applications, no. But it ruins some calculus proofs.

For that reason, calculus distinguishes various degrees of smoothness. Differentiable means that f' exists throughout the interior of an interval, twice differentiable means that f'' exists throughout, and so on. Smooth in calculus jargon means that we can keep differentiating it forever.

4.2. ACCELERATION

Gravity

In Chapter 2.2, we calculated net displacement by integrating velocity and calculated velocity by integrating something else. That something else is known as acceleration. Acceleration is the first derivative of velocity with respect to time and hence the second derivative of distance. The only case we considered was constant acceleration a. Given initial velocity $v(0) = b$, the first integral of a tells us that $v(t) = at + b$. The second integral tells us that distance $y(t) = \frac{1}{2}at^2 + bt$ given initial position $y(0) = c$. This is the equation of a parabola.

This double integral has generated a host of applications thanks to gravity and artillery. Gravity supplies an earthly force of near-constant acceleration. Artillery supplies projectiles that obey gravity. The applications help projectiles reach their targets.

The connection between gravity and acceleration took ages to figure out. The famous philosopher Aristotle thought that each material had a natural home that it was drawn to—heaven for smoke, earth for rock—and that once home it tended to stay put. He also thought that heavy objects fall faster than light ones. For many centuries this was so widely believed that few people bothered to test it.

Four hundred years ago, Galileo Galilei pioneered the notion of inertia, which means that motion persists unless something else intervenes. And while balls fall faster than feathers, that's because air interferes in ways that gives feathers lift. When Galileo let heavy and light balls roll down slightly inclined planes, they moved the same.

Furthermore, their acceleration was constant. Its value, usually denoted g, is about 9.8 meters per second per second (m/s²). Ignoring wind gusts and air friction, we can use g to answer various questions:

Q: *If we drop a ball from a platform h meters high, how many seconds t does it take to hit ground?*

A: Since $h = \frac{1}{2} g t^2$, $t = \sqrt{2h/g}$.

Q: *How fast is the ball dropping just before it hits ground?*

A: $v = gt = \sqrt{2gh}$.

Q: *If we throw a ball straight up from the ground with the speed $\sqrt{2gh}$ it dropped, how high does it go and how long until it gets there?*

CHAPTER 4

A: The way up is like a rewind of the previous ball's fall, so the height reached is h and the time needed to get there is $\sqrt{2h/g}$.

Q: *If we throw a ball straight up from the ground with speed b, how high does it go and how long until it returns to ground?*

A: Rearrange $b = \sqrt{2hg}$ to see that $h = \tfrac{1}{2}b^2/g$. Since the ball takes b/g seconds to stop moving up and b/g seconds to fall back down, total time is $2b/g$.

Artillery

From a combatant's perspective, the previous formulas might seem worse than useless. The first rule of cannons is never to point them straight up. The GeoGebra activity "Cannon Practice" lets you aim somewhere else.

While many different trajectories get to where we want and all of them mimic parabolas, we need to tune the speed and angle. Basketball players make similar choices when they shoot balls toward the hoop, with their own motion adding extra complexity. While they don't consciously pose equations, years of practice help them learn intuitive solutions.

Cannon fire is easier than basketball to analyze mathematically. The explosive charge and projectile weight determine the initial velocity b. The horizontal component b_x and vertical component b_y satisfy $b^2 = b_x^2 + b_y^2$. The cannon angle determines the slope $s = b_y/b_x$.

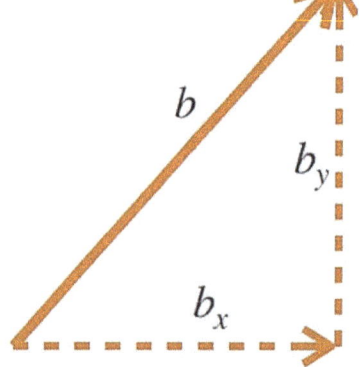

Gravity has no influence on horizontal velocity. This reflects a general principle known as Newton's Second Law of Motion. Acceleration a equals force F divided by mass and has the same direction as F. Hence $x = b_x t$ and $y = -\frac{1}{2}g t^2 + b_y t + h$, apart from air friction which we will ignore for now. Substituting x/b_x for t and $b^2/(1+s^2)$ for b_x^2 shows that $y = -\frac{(1+s^2)g}{2b^2} x^2 + sx + h$. Let's use this to solve more problems:

Q: *If $h = 0$, where does the cannonball land?*

A: At $y = h = 0$, $\frac{(1+s^2)g}{2b^2} x^2 = sx$, with solutions of $x = 0$ for launch and $x = \frac{2sb^2}{(1+s^2)g}$ for landing.

Q: *How should we aim the cannon to maximize x?*

A: Choose $s > 0$ to maximize $\frac{s}{1+s^2}$ or minimize $\frac{1+s^2}{s} = \frac{1}{s} + s$. The latter has derivative $1 - s^{-2}$ and second derivative $2s^{-3} > 0$, so the solution is $s = 1$. Shoot the cannon at an angle of $45°$.

Q: *What is the maximum x and corresponding t?*

A: Substitute $s = 1$ to see that $x = \frac{b^2}{g}$ and $t = \frac{b^2}{g\,x} = \frac{b\sqrt{2}}{g}$.

Pesky Aliens

Imagine that annoying space aliens drop jellybeans on us from zillions of miles away. Each jellybean is specially coated to deflect all forces other than Earth's gravity. How fast v_{crash} does each jellybean crash into Earth?

CHAPTER 4

The previous formula won't work because gravity declines with distance squared. Two spherical masses M and m with centers r apart attract each other with force $-GMm/r^2$, where G is a universal constant and the negative sign draws the masses closer. For a jellybean dropped gently at time 0 toward Earth with mass M_e, its velocity at time T is

$$v(T) = \int_0^T \frac{-GM_e}{r^2(t)} dt$$

This integral can't be calculated directly since $r(t)$ is an integral of v. Instead we turn to a powerful indirect method. Define work W as the integral of force $F(r)$ over the distance dr it is applied. Substituting $F = ma = mv'$ and $dr = r'dt = v\,dt$,

$$W(T) = \int_{r(0)}^{r(T)} F\,dr = \int_0^T mv \frac{dv}{dt} dt = \int_{v(0)}^{v(T)} mv\,dv = \tfrac{1}{2}mv^2(T) - \tfrac{1}{2}mv^2(0).$$

The expression $\tfrac{1}{2}mv^2$ is known as kinetic energy. Hence work equals the change in kinetic energy. For each jellybean hitting Earth at radius R given $v(0) = 0$,

$$\tfrac{1}{2}mv_{crash}^2 = W = \int_\infty^R \frac{-GM_e m}{r^2} dr = \left.\frac{GM_e m}{r}\right|_\infty^R = \frac{GM_e m}{R}.$$

which implies $v_{crash} = -\sqrt{2GM_e/R}$.

Great! Compute the square root and we're done. Only first let's make the numerator more inviting. On the surface of Earth, $GM_e/R^2 = g$, which implies $v_{crash} = -\sqrt{2gR}$. Earth's circumference is $40{,}000$ kilometers (km), which implies $R \approx 40{,}000/(2\pi) \approx 6{,}400$ km. It follows that each jellybean strikes Earth at velocity $\sqrt{2 \cdot (9.81/1000) \cdot 6400} \approx 11.2$ km per second (km/s). At a mass of one gram, each impact is equivalent to a medium-sized truck crashing at 90 km per hour. Watch out!

Now for the good news. Our Sun is 330 thousand times more massive than Earth and a mere 150 million km away. It is big enough and close enough to lure most space debris away. Huge Jupiter adds protection on the side away from the Sun. The unusual combination provides a calm oasis relative to most planets in the universe.

Still, it is fun to explore firsthand if we can manage it. How fast do we need to blast off to escape Earth's gravity? The answer, called escape velocity, is just $-v_{crash}$. Here is a table of escape velocities from various surfaces.

Surface Location	Escape Velocity (km/s)	Surface Location	Escape Velocity (km/s)
Moon	2.4	Uranus	21.3
Mercury	4.3	Neptune	23.8
Mars	5.0	Saturn	35.6
Venus	10.3	Jupiter	59.6
Earth	11.2	Sun	618

The big numbers raise a host of questions about how to fuel takeoffs and landings. One answer is to swing with gravity and not just fight it. Since the rate of change of kinetic energy is $\left(\frac{1}{2}mv^2\right)' = mvv' = mva = Fv$, waiting for gravity to speed up the craft before burning fuel will transfer more kinetic energy from exhaust gas to craft. This is known as the Oberth effect, after the father of modern rocketry who first described it.

4.3. CENTRIPETAL FORCE

Hammer Throw

It's hard to throw heavy things far. Your force on the object vanishes once it leaves your hands and its acceleration divides force by weight. For better result, attach a handle and twirl the object first. Twirling allows more time to apply force before you release.

Ancient Celts turned this notion into a mix of weapon and entertainment, with sledgehammers the eventual object of choice. Modern technology replaced the head with metal ball and the handle with wire and grip. Modern culture added rules and made it a bloodless sport. Now called the hammer throw, it has been played in every Olympics since 1900.

CHAPTER 4

Watching a hammer thrower in action, we see that:

- It takes strong legs to speed up the twirl;
- It takes strong arms to keep the hammer from breaking free;
- The hammer flies straight regardless of the twirling.

These are all illustrations of Newton's Laws applied to motion in a circle. Force is needed to keep turning the hammer in a circle. The force is directed inward, through the pull of arms and wire. Once released, the hammer moves along its tangent line.

Let's quantify this. Suppose a small object moves in a circle of radius r with constant speed v. The horizontal velocity is $v_x = x'$ and the vertical velocity is $v_y = y'$. Since $v_x^2 + v_y^2 = v^2$, (v_x, v_y) lies on a circle of radius v.

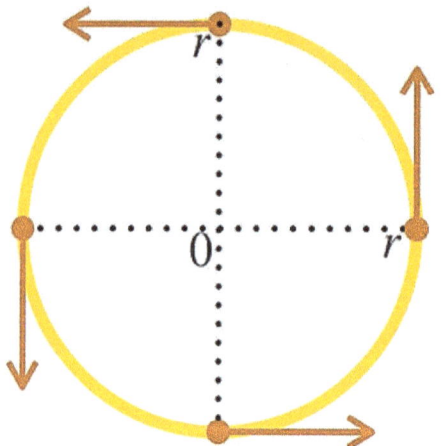

By the chain rule, $v_y = \frac{dy}{dx} v_x$, so v matches the slope $v_x/v_y = -x/y$ of the tangent line and points perpendicular to a radial line from the center. For counterclockwise motion, $(v_x, v_y) = (-yv/r, xv/r)$. For clockwise motion, reverse the signs on v_x and v_y.

In other words, the first derivative maps circles to circles, with radius multiplied by v/r and position rotated 90° forward. Since acceleration is the first derivative of velocity, it is circular too, with yet another multiplication of radius by v/r combined with a rotation 90° forward. Compared with the original position, acceleration flips the signs and multiplies by v^2/r^2. That is, $(x'', y'') = (-xv^2/r^2, -yv^2/r^2)$. This confirms that acceleration points directly back to center. It is called centripetal force and has magnitude

$$\frac{v^2}{r^2}\sqrt{x^2 + y^2} = \frac{v^2}{r}.$$

Falling Around

Imagine a rock of mass m whizzing through space with no force affecting it other than gravity from a sun of mass M a distance r away. Rock accelerates toward sun at rate GM/r^2 while sun accelerates toward rock at rate Gm/r^2. Total net acceleration is $G(M+m)/r^2$. Typically M is so much larger than m (written $M \gg m$) that we can ignore rock's pull on sun, like we ignored jellybean pull on Earth.

Suppose that at some moment the rock moves perpendicular to the sun with speed $v = \sqrt{GM/r}$. In that case, $v^2/r = GM/r^2$. Gravity will supply exactly the centripetal force needed to keep both r and v constant. Rock will fall around sun for eons in a circle.

Planets don't fall into the sun either. As the conditions above aren't fully met, orbits are ellipses rather than circles. However, since the ellipses aren't that squashed and we're not yet ready to tackle them, let's stick with circular approximations. How might we test the theory?

A simple test checks for consistency across the various $v^2 r$. Since the distance vT traveled in one orbital period T equals $2\pi r$, the relation $GM = v^2 r$ implies $T^2/r^3 = 4\pi^2/(GM)$. The period squared should be proportional to the radius cubed.

This relationship is the core of astronomer Johannes Kepler's Third Law, first published in 1619. The table to the right presents modern estimates, where distance is the geometric mean of minimum and maximum distances.

Planet	Period In Earth Years	Relative Distance from Sun
Mercury	0.24	0.39
Venus	0.62	0.72
Earth	1.00	1.00
Mars	1.88	1.52
Jupiter	11.9	5.21
Saturn	29.5	9.54
Uranus	84.0	19.2
Neptune	165	30.1

CHAPTER 4

Kepler's Laws thrilled astronomers. These connected a host of observations to some master design: "music of the spheres" as Kepler called it. When Newton used calculus to trace the Laws to gravity, it was a great victory for scientific analysis generally.

Similar calculations apply to orbits around Earth. In 1945 they inspired an amazing proposal from science fiction writer Arthur Clarke. How about using not-yet-invented space rockets to launch not-yet-invented satellites into stationary orbits to relay communications from Earth?

By stationary Clarke meant travel eastward along the equator with a period of exactly one day, so that the satellite looks still from the ground. Similar ideas had been propounded by Herman Potočnik in 1928 and Konstantin Tsiolkovsky in 1895. However, Clarke captured public imagination for something on the verge of being feasible. In 1964, a satellite entered geostationary orbit for the first time. Hundreds now orbit in a narrow band above the equator known as the Clarke Belt.

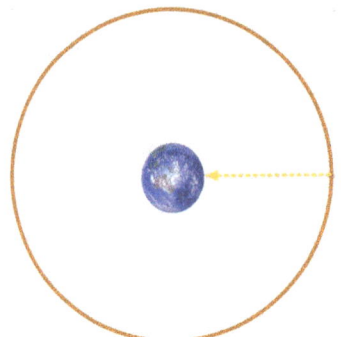

Orbits far from the equator cannot appear stationary, as they must extend as far south as they do north. However, many satellites have geosynchronous orbits, meaning that they return daily to the same spot.

A Clarke orbit has radius $r^* = \sqrt[3]{GM_e D^2/(4\pi^2)}$ where D is the time the Earth needs to rotate around its axis. In one year, the Earth rotates one more time than we notice since it revolves once around the Sun. Hence one rotation, called a sidereal day, takes 24 hours times $365.26/366.26$ or $86,164$ seconds. From the previous section, $GM_e = gR^2$, where Earth's surface gravity and radius can be estimated more precisely as $g \approx 9.807$ m/s² and $R \approx 6,375$ km. This implies $r^* \approx 42,160$ km or $35,780$ km above ground. While a satellite in Clarke orbit seems to stand still, it actually revolves at 3.07 km/s.

Tides

The Sun pulls Earth from $r = 150$ million km with average acceleration $a_{sun} = GM_{sun}/r^2$ of 6 millimeters per second squared. Since distance varies from $r - R$ to $r + R$, the maximum variation in solar acceleration is approximately

$$\left(-\frac{GM_{sun}}{r^2}\right)' \cdot 2R = \frac{2GM_{sun}}{r^3} \cdot 2R = a_{sun} \cdot \frac{4R}{r} \approx \frac{a_{sun}}{6000}.$$

Earth's pull on us is 10 million times stronger!

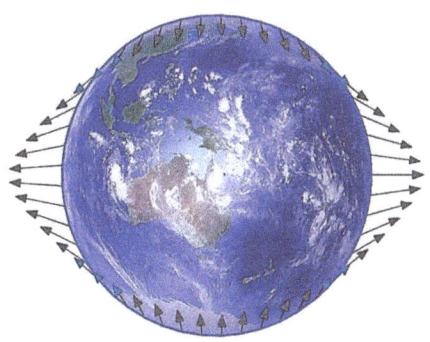

Do those tiny fluctuations matter? For tectonic plates, not much, but for oceans they create two bulges. One is closest to Sun, where water pulls away most from Earth. The other is farthest from Sun, where Earth pulls away most from water. As peaks shift by half a kilometer per second, water chases after and builds into bigger waves called tides.

The corresponding range of lunar gravity is $4Ra_{moon}/r_{moon}$. While $a_{sun} \approx 175 a_{mn}$ thanks to Sun's 27-million-times advantage in mass, $r_{sun} \approx 394 r_{moon}$. This makes lunar gravity vary $394/175 \approx 2.25$ times more than solar gravity does. Hence, lunar tides are what we most notice.

Since the Moon orbits Earth in the same direction that Earth rotates, Earth needs slightly more than a day to catch up with Moon. Also, Moon's orbital period of 27.3 days looks like 29.5 days from a vantage point on Earth orbiting Sun. As a result, the near-daily cycle of two high tides and two low tides lasts 30.5/29.5 days or 24 hours and 50 minutes. Solar influence is strongest when Sun, Earth and Moon line up nearly straight. On Earth we see that as new moon or full moon, so even then Moon seems to dominate.

Centrifuges

For things pulled by gravity to stay put, something needs to push back. That pushback is what we feel as weight. We don't feel Earth falling around Sun because we're all falling together. Sky divers don't feel weight until their parachutes open. Yet jet planes taking off make their passengers feel heavier because the accelerating seats push harder. The ratio of apparent gravity to normal surface gravity is called the *g*-force.

We can all breathe with relief about normal gravity as otherwise we couldn't breathe. Normal gravity keeps air denser at sea level than at mountain tops, but not so dense that we can't live on mountains or walk upright. Normal gravity also helps dense stuff settle below less-dense stuff, like rock and oceans below air, without insisting that our bones settle below our organs.

When *g*-force increases, it presses denser stuff to separate more from less-dense stuff. How good or bad that is depends on the context. For example, our bodies handle surface gravity $1g$ a lot better than either $0g$ (where our muscles quickly atrophy) or $5g$ (where our hearts can't maintain an even flow of blood).

No cars or trucks can achieve $3g$ for long; they would need to speed up every second by 110 kilometers per hour. However, rotating containers called centrifuges manage this with relative ease. The NASA centrifuge below can maintain $20g$.

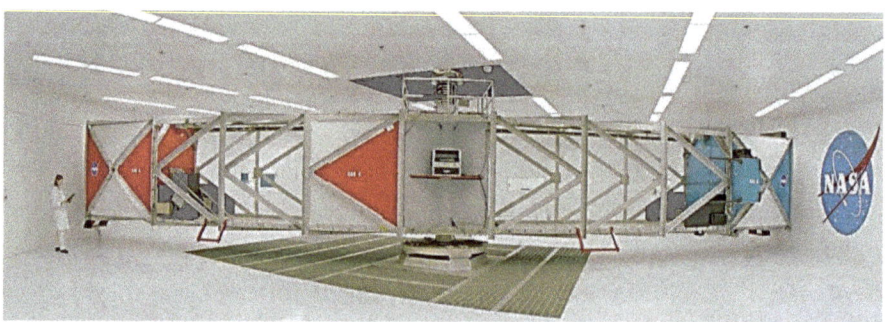

In contrast to gravity, centrifuges drive the densest stuff away from the center. The centrifugal force seems to work like antigravity. In reality the centrifuge is pulling its own wall toward the center, and the centrifugal force is the effective weight of stuff continuing on a tangent line and pushing back.

Given centripetal acceleration $a = v^2/r$, let $H = a/g$ denote the g-force. Since orbital period $T = 2\pi r/v$, $H = 4\pi^2 r/(gT^2)$. When measuring in meters and seconds, $\pi^2/g \cong 1.003$, so $H \approx 4r/T^2$. A wheel-shaped space station one kilometer in diameter would need to rotate about once per minute to generate g-force of 0.5. For a given kinetic energy $\tfrac{1}{2}mv^2 = \tfrac{1}{2}mgrH$, the most efficient way to generate high H is to keep r small. A washing machine centrifuge 0.5 meters wide can rotate up to 20 times per second, generating a maximum $400g$. A laboratory centrifuge 0.1 meters wide can rotate up to 200 times per second, generating a maximum $8000g$.

Centrifuges have many applications, including:

- Training fighter jet pilots to handle tight turns
- Training astronauts to handle reentry
- Helping amusement park visitors "defy" gravity
- Testing building materials' resistance to earthquakes
- Wringing water out of wet clothes
- Removing waste products from water and drilling fluids
- Separating cream from milk
- Refining and crystallizing sugar
- Extracting olive oil from crushed olives
- Collecting enriched uranium for nuclear reactors
- Measuring the concentration of red blood cells
- Isolating DNA from other cell material.

CHAPTER 4

4.4. CURVE DESIGN

Banked Turns

So far we've looked at two ways of supplying centripetal force. In one, gravity pulls an object moving sideways. The other attaches arms or walls to a rotating center. Yet many things make turns around vacant centers. They have to supply the centripetal force themselves. How?

The most direct approach pushes at right angles to the line of motion. Running in circles, we turn our feet slightly inward and press harder on the side closer to the center. Cleats or spikes make it easier to apply lateral force without losing traction.

Fast runners do something extra: they tilt toward center, or "lean into turns" as they call it. Tilts are even more dramatic in sports with faster motion (higher v) and sharper turns (smaller r).

104

Leaning requires extra force to maintain balance, which is why we learn to walk upright without leaning. Something in that extra force must boost centripetal acceleration $a = v^2/r$ or racers wouldn't apply it.

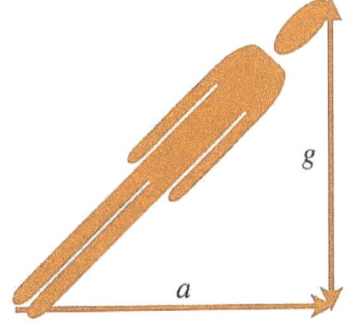

That something is its horizontal component. Recall that forces have direction as well as magnitude. Racer tilt redirects force upward and inward.

The upward force must match gravity mg to keep a racer of mass m from dropping or rising. The inward component ma supplies the centripetal force. The racer's slope s equals g/a. Hence a will match v^2/r when $s = g/v^2$. The tighter the turning radius or higher the velocity, the more the racer must tilt to reduce s.

As cars don't readily tilt, they need thick tires to provide lateral traction when turning. That's not always enough. To make tight curves safer at high speed, engineers build embankments that tilt inward. Banked turns accelerate toward the center while reducing lateral stress.

CHAPTER 4

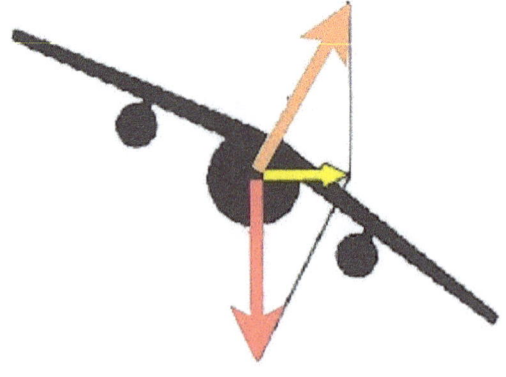

Airplanes too bank turns. They have to, as air provides far less lateral traction than ground does. The difference between the pressure under a curved wing and the pressure above the wing creates lift. The airplane's tilt directs lift inward.

Curvature

In determining safe combinations of velocity and slope, the main challenge is to figure out the relevant radius r. Roads or flight paths hardly ever make perfect circles. However, if a curve is smooth enough to have second derivatives, we can view it as a continuous progression of tangent circles instead of just tangent lines. Each tangent circle should match the curve in both first and second derivatives.

A tangent circle is also known as an osculating circle. Its size is very sensitive to curve shape as the GeoGebra activity "Osculating Circle" shows.

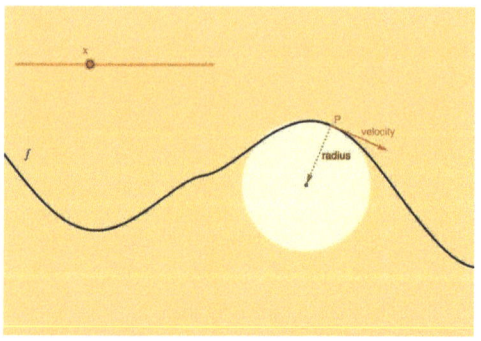

If we graph a function f, we can fit a tangent circle at any point and measure its radius. However, this is tedious and prone to error. Calculus lets us compute radii without graphing circles first. For travel along f with horizontal velocity $v_x = \dfrac{dx}{dt} = 1$ and vertical velocity $v_y = \dfrac{df}{dx} \cdot \dfrac{dx}{dt} = f'$, the Pythagorean theorem indicates a total speed of $v = \sqrt{1 + (f')^2}$. Apply the chain rule again to calculate

106

$$a = \frac{dv}{dt} = \frac{1}{2\sqrt{1+(f')^2}} \cdot 2\frac{df'}{dx} \cdot \frac{dx}{dt} = \frac{f''}{\sqrt{1+(f')^2}}.$$

Rewriting our earlier $a = v^2/r$ result for circles yields a formula that looks intimidating but is easy to apply:

$$r = \frac{v^2}{a} = \frac{\left(1+(f')^2\right)^{3/2}}{f''}.$$

Roller Coasters

Roller coasters are founded on the principle that acceleration can be scarily fun. A particularly fun scare takes riders through a vertical loop. The core challenge for the roller coaster designer is to ensure that g-forces are high enough to stay on track without being so high that they harm riders. Let's use calculus to analyze this.

Recalling the relation between force and energy, a coaster that falls h gains kinetic energy $\tfrac{1}{2}mv^2$ equal to work $\int_0^h mg\,dx = mgh$ done by gravity, which implies $v^2 = 2gh$. If the coaster is upside down at the top of a circle of radius r, it will fall off the tracks unless $v^2/r > g$. Hence the top of the loop must be at least $h > \tfrac{1}{2}r$ lower than the coaster's starting point.

Unfortunately, the solution creates another problem: jerks. Jerk is the apt name for the first derivative of acceleration, which is also the second derivative of velocity and the third derivative of distance traveled. Big jerks can be dangerous. Our necks are particularly sensitive and prone to whiplash. To reduce injury, designers increase r at the bottom of loops to dampen the extra g-forces there and ease transitions into and out of sharp curves. As a result, roller coaster loops more closely resemble spirals or inverted tear drops than circles.

Roller coaster designers could be more daring if riders had more structural padding. That's how Mother Nature dealt with a related challenge. Horses fend off predators using powerful bucks and kicks. These are big jerks, which rattle the jerker too. To keep their wits, horses gained extremely strong necks.

In everyday road and rail transport, we're not aiming for either our thrills or others' spills. Making the curvature $1/r$ linear in distance traveled can help reduce jerks at major interchanges. This creates a fascinating path known as a clothoid, spiros, Euler spiral or Cornu spiral. It is embedded in tens of thousands of off-ramps, on-ramps and roller coaster loops.

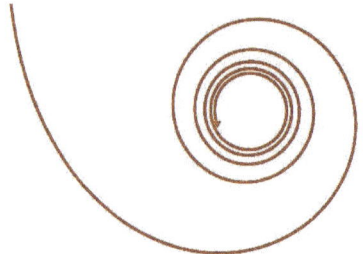

The formula for a clothoid is an integral that can't be solved using familiar functions. We won't tackle it here; even professionals find it challenging. Watch what it does in the GeoGebra activity "Cornu Spiral".

Cubic Spline

Drawing straight lines and circles is easy. Drawing any other curve can be hard. One method plots key points as dots, connects the dots with straight lines, and then bends the lines to soften corners. A century ago, engineers would trace splines—elastic strips of wood or rubber—fastened between key points. As the splines bend the least they can to make those connections, this generates relatively smooth curves.

Nowadays, computers do the softening for us. Above is a chart of two zigzags. Both bounce between −1 and +1 moving one unit at a time. The only difference is how they bounce: straight or curved.

The curve illustrates a cubic spline. Given $n+1$ ordered points numbered from 0 to n, the cubic reference indicates that spline i takes the form $a_i x^3 + b_i x^2 + c_i x + d_i$ between points $i-1$ and i. By defining $t = (x - x_{i-1})/(x_i - x_{i-1})$, we can rewrite the spline as

$$C_i(t) = (1-t)y_{i-1} + ty_i + t(1-t)\left((1-t)g_i + th_i\right)$$

for t between 0 and 1. Its first derivatives at the edges are $C_i'(0) = y_i - y_{i-1} + g_i$ and $C_i'(1) = y_i - y_{i-1} - h_i$. Its second derivatives at the edges are $C_i''(0) = -4g_i + 2h_i$, and $C_i''(1) = 2g_i - 4h_i$.

Splines match in slope where they meet, so $C_i'(1) = C_{i+1}'(0)$ and $g_{i+1} = -y_{i+1} + 2y_i - y_{i-1} - h_i$ for $i=1$ to $n-1$. If we make the outer segments straight, like in the illustration, then $g_1 = \tfrac{1}{2}h_1$ and $h_n = \tfrac{1}{2}g_n = \tfrac{1}{2}(-y_n + 2y_{n-1} - y_{n-2} - h_{n-1})$. The remaining coefficients h_1 through h_{n-1} can be determined in either of two ways:

- Make the slope s_i at interior point i match the slope of the secant from point $i-1$ to point $i+1$, in which case $h_i = y_i - y_{i-1} - s_i$. This is how most computer graphics smooth animated motion; the illustration used that method too.

- Make the second derivatives match where splines meet, so $C_i''(1) = C_{i+1}''(0)$. This implies a linear relation between h_{i-1}, h_i, and h_{i+1}. While the resulting chain of equations is slightly harder to solve, it makes the curves a bit less wriggly.

Bézier Curves

Cubic splines also assist computer art, where we're less concerned about fitting specific interior points than creating attractive curves. Standard curve-drawing tools use control points to pull and push curves around. These curves are called Bézier and are constructed recursively:

- A linear Bézier B_{12} is a linear interpolation between two points P_1 and P_2. Its formula is $B_{12}(t)=(1-t)P_1+tP_2$ where t varies from 0 to 1.
- A quadratic Bézier B_{123} is a linear interpolation between two linear Bézier B_{12} and B_{23}, with formula $B_{123}(t)=(1-t)B_{12}(t)+tB_{23}(t)$.
- A cubic Bézier B_{1234} is a linear interpolation between two quadratic Bézier B_{123} and B_{234}, with formula $B_{1234}(t)=(1-t)B_{123}(t)+tB_{234}(t)$.

We can keep adding layers and some people do. However, cubic Bézier usually provides enough flexibility, with two control points for the curvature of each segment. To use them, artists never need to jot down numbers or formulas; they just move the control points by hand.

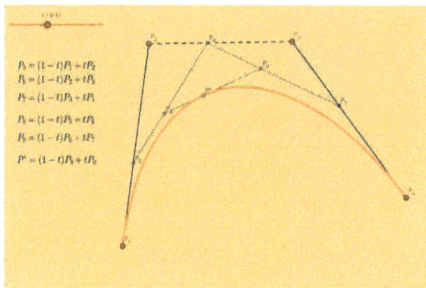

The GeoGebra activity "Cubic Bézier Scaffolding" shows how this works. Normally all the user sees are the curve and the four connected control points. This version reveals more scaffolding to show how the curve is constructed.

Are you too feeling pulled and pushed in too many directions to keep them straight? Don't worry. It takes practice to get comfortable with slopes of slopes. Just appreciate how useful second derivatives are, whether you're shooting down alien jellybeans or designing roller coasters. The next chapter will extend derivatives to slopes of slopes of slopes of....

5

Iteration

Calculus shows how to refine approximations through iteration. The algorithms simplify and speed computations. Moreover, the infinite series they imply often reveal a lot about functions' behavior.

5.1. APPROXIMATION

Wonderful Calculators

There are only so many digits we can handle before we run out of storage space or steam. Missing bits act like infinitesimals. Usually they drop out without messing things up. However, we must be careful when dividing by small values or subtracting two nearly equal values. Also, since small errors can build into big ones, it is best to compute each step to extra precision and round excess digits at the end.

For example, the calculator app on my computer rounds $\sqrt{2}$ to 1.414213562373095. When I square that back, it displays 2 exactly. Is the calculator lying, or can it not tell the difference given the number of digits it

x	x^2
1.414213562373093	1.99999999999999
1.414213562373094	2.00000000000000
1.414213562373095	2.00000000000000
1.414213562373096	2.00000000000000
1.414213562373097	2.00000000000001

111

CHAPTER 5

stores? To check, I nudge the last digit in the expansion. Nudge by 1 and nothing changes. Nudge by 2 and the square gets 10^{-15} higher or lower. The calculator is trying the best it can.

How much precision do we need? In theoretical analysis, all we can get. Not only does $\sqrt{2}$ point to its origins, but also it might cancel out another $\sqrt{2}$ through division or fuse through multiplication into 2. In applied work, precision needs vary with what we want to do and how skilled we are at doing it. If I want a fence to enclose 2 square meters, I'm happy if each side measures 1.41 meters. For machines to run efficiently, precision must be much higher.

The quest for speedy high-precision computation stretches back hundreds of years. In 1614 John Napier published a long table of values for a new function he named logarithm. He called the table "Wonderful", which it was, as it converted tedious multiplications and divisions into much simpler additions and subtractions. Twenty years of hard work enabled Napier to provide 7-digit accuracy, to the delight of users.

Nowadays any scientific calculator can instantly generate a million-fold higher precision. We tend to take it for granted. Yet it's one of the marvels of the modern world. How does the calculator do it? If you think it stores square root tables for every 15 decimal digit input, guess again. The inputs alone would require thousands of times more storage than most computers provide. Besides, it isn't efficient. To see why, engage a friend in competition. One of you multiplies by hand two 7-digit numbers. The other searches a library for the seventh book on the third shelf of the sixth row and writes down the twentieth word in the fifth chapter. Who finishes first? I'll bet on the multiplier. Search is hard!

Computers don't like lookup either. They can compute n digits of complicated functions far faster than they can search through 10^n entries, and the gap scales up with n. Hence they keep their lookup tables short. To speed computation, they apply mathematical recipes known as algorithms.

Easy Square Roots

The oldest algorithm for square roots dates back to ancient Babylonia. It is extremely easy to apply and analyze. Moreover, the analysis lends itself to calculus-inspired generalization.

Given a number Y, let x denote an estimate of $\sqrt{Y}$. To check how good it is, let's compare x to Y/x. If they are so close we can't tell them apart, we have our answer. Otherwise, let's take their average $\frac{1}{2}(x+Y/x)$ as a new estimate of $\sqrt{Y}$ and repeat until we're either happy or grossly unhappy.

For example, suppose $Y=2$ and the initial $x=1$. We average their ratio 2 with x to generate a new estimate 1.5. Using $x=1.5$, $Y/x=1.333...$, which generates a new estimate $1.4166....$ In two rounds we've matched the first three digits of $\sqrt{2}$. The third round estimates $1.41421...$, which is accurate to six digits. The fourth-round estimate is accurate to twelve digits. By the fifth round we have outdone the calculator.

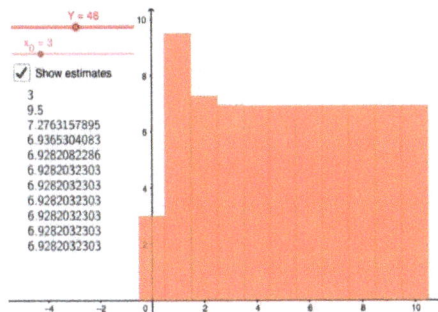

To reassure yourself this isn't a fluke, try the GeoGebra activity "Easy Square Roots". It usually converges on the correct answer within five rounds. As a buffer against bad guesses, the GeoGebra activity adds a few more rounds than typically needed.

Still, there are ways to sabotage it. My personal favorite is to plug in a negative Y, which has no real number solution. Here are the first 50 estimates of $\sqrt{-2}$ given an initial estimate of 1.

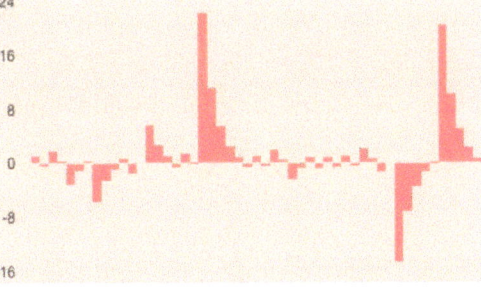

113

CHAPTER 5

If x is huge, Y/x is tiny, so the next estimates roughly halve until they get closer to $|Y|/x$. When $Y<0$, any x that gets very close to $|Y|/x$ makes their average tiny, in which case the next estimate explodes. Yet there is no absolute regularity. A small change in the initial x can drastically alter the timing and intensity of explosions.

While setting $Y=0$ doesn't breed chaos, it gums up the works in a different way. For any $x \neq 0$, $Y/x = 0$, so the next estimate of x just halves it and only slowly rounds to 0. On reflection, any positive Y can generate similar behavior if the initial x differs by many orders of magnitude from the true $\sqrt{Y}$.

The best defense is good seeding. If we rewrite Y as $V \cdot 10^{2P}$, where $0.1 < V < 10$ and P is an integer, a seed of 10^P will converge within six rounds. Since computers store numbers in base 2, they can easily seed x to within half or twice $\sqrt{Y}$. Moreover, to halve a sum they just need to shift a digit. This makes the algorithm even simpler for computers than for us, and not terribly slower than their best square root method.

To see how this works, let's track the relative error $\delta = x/\sqrt{Y} - 1$. Since $x = (1+\delta)\sqrt{Y}$, the next estimate is

$$\tfrac{1}{2}\left(1+\delta+\frac{1}{1+\delta}\right)\sqrt{Y} = \frac{2+2\delta+\delta^2}{2(1+\delta)}\sqrt{Y} = \left(1+\frac{\delta^2}{2(1+\delta)}\right)\sqrt{Y}.$$

Hence the relative error shrinks to $\tfrac{1}{2}\delta^2/(1+\delta)$, which for small δ is close to $\tfrac{1}{2}\delta^2$. Since an error of 10^{-k} signifies roughly k-digit accuracy in decimal expansions, the number of matching digits tends to double at every round. The formula also illuminates the vulnerabilities to sabotage. If Y is negative, δ isn't a number that can shrink to zero through squaring. If $x=0$, $\delta = -1$, so the $1+\delta$ in the denominator blows up. If $x=\infty$, $\delta = \infty$ and never gets finite again.

Finding Zeros

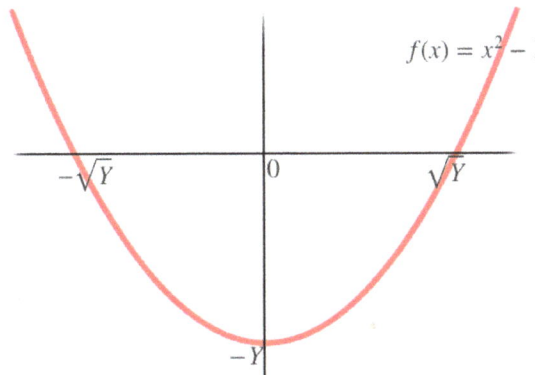

Here is another way to think about square roots. Consider the function $f(x) = x^2 - Y$. It equals zero at $\sqrt{Y}$ and $-\sqrt{Y}$. Hence the square root problem amounts to finding a zero of f.

Imagine the graph of f marks a trail that we're searching for a target altitude of zero. Starting from $(x_0, f(x_0))$, each round n moves us to position $(x_n, f(x_n))$. Unfortunately, fog obscures the trail. How can we use current position and slope to help find a zero?

Suppose we project the current tangent line to the x-axis and sets x_{n+1} where it intersects. Since we have to stay on the trail, our altitude will be $f(x_{n+1})$ rather than 0. Still, the approximation will likely improve and we can use it as starting point for the next estimate. The process is known as iteration. Every round draws on the results before.

To extend from $(x_n, x_n^2 - Y)$ to $(x_{n+1}, 0)$ with slope $2x_n$, we must have

$$\frac{Y - x_n^2}{x_{n+1} - x_n} = 2x_n,$$

which implies

$$x_{n+1} = x_n + (Y - x_n^2)/2x_n$$
$$= \tfrac{1}{2}(x_n + Y/x_n).$$

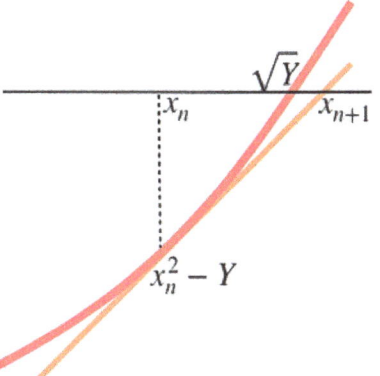

That's the easy square root algorithm!

CHAPTER 5

Newton-Raphson Method

We can easily generalize this to search for zeros of any differentiable f. Equating the tangent slope $f'(x_n)$ to the desired rise $-f(x_n)$ over the run $x_{n+1} - x_n$ implies the update

$$x_{n+1} = x_n - \frac{f(x_n)}{f'(x_n)}.$$

This is known as the Newton-Raphson method. It is easy to use and works so well close to a zero that more complicated algorithms often switch to Newton-Raphson for fine-tuning. The main challenge is getting close. For example, consider the power function $f(x) = x^m - Y$ with a zero at $\sqrt[m]{Y}$ for $Y > 0$. Its Newton-Raphson update simplifies to

$$x_{n+1} = x_n + \frac{Y - x_n^m}{mx_n^{m-1}} = x_n \left(\frac{m-1}{m} + \frac{1}{m} \cdot \frac{Y}{x_n^m} \right).$$

When m is positive, this is guaranteed to converge, although the pace will be slow for a poor seed. When m is negative, a seed that is too high will cause estimates to diverge to $-\infty$.

The case $m = -2$ has found use in videogames. To rescale objects, they often divide by square roots of sums Y of squares, or equivalently multiply by $Y^{-1/2}$. Faster calculation of $Y^{-1/2}$ noticeably speeded *Quake*, a popular video game in the late 1990s. It used an ingenious seed x_0 with an error of less than 4% and then applied the Newton-Raphson update $x_1 = x_0(1.5 - 0.5Yx_0^2)$ to trim the error to 0.15%.

The case $m = -1$ is even more interesting. To compute $x = 1/Y$ without division, we can apply Newton-Raphson to $x^{-1} - Y$. The update is $x_{n+1} = x_n(2 - Yx_n)$. Repeating it is usually faster than standard division since it doesn't require trial and error. To speed convergence, rewrite Y in power-of-two form as $(1-y)2^P$ with $0 < y < \frac{1}{2}$ and set $x_0 = 2^{-P}$. Our first update works out to $x_1 = x_0(1+y)$, our second to $x_2 = x_1(1+y^2)$, and our third to $x_3 = x_2(1+y^4)$. At every iteration n, we compute

$$x_{n+1} = x_n(1+y_n) \text{ and } y_{n+1} = y_n^2.$$

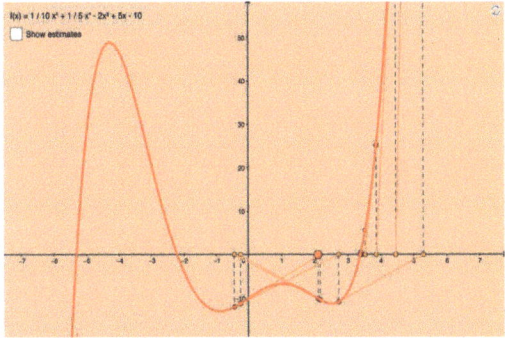

The GeoGebra activity "Newton-Raphson Method" can be used with any smooth f and any seed x_0. While it usually finds a zero, it doesn't always find the closest zero. Occasionally it oscillates between nonzeros.

5.2. TAYLOR SERIES

Higher-Order Approximation

Newton-Raphson approximates a smooth $f(x)$ with a tangent line $T_1(x) = f(a) + f'(a)(x-a)$ at the current estimate a, which I renamed from x_n to reduce clutter. As we saw with tangent circles, we can improve the fit by matching the second derivative too. The simplest match is a parabola with formula $T_2(x) = T_1(x) + \frac{1}{2} f''(a)(x-a)^2$. If we set $T_2(x) = 0$, we can rearrange into

$$x - a = \frac{-f}{f' + \frac{1}{2}(x-a)f''} \approx \frac{-2ff'}{2(f')^2 - ff''},$$

where the last step replaces the second $x-a$ with the Newton-Raphson nudge $-f/f'$. All terms in f are evaluated at a. This is known as Halley's method. Two Halley iterations usually outperform three Newton-Raphson iterations but require more computation.

Sometimes it is easier to fit one long polynomial than to iterate shorter polynomials. Given $T_0 = f^{(0)} = f$ and $0! = 1$, recursively define $T_m(x) = T_{m-1}(x) + t_m(x)$, where $t_m(x) = f^{(m)}(a)(x-a)^m / m!$, $f^{(m)} = \left(f^{(m-1)}\right)'$, and the "factorial" $m! = m \cdot (m-1)!$. To express this more neatly, write

$$T_m(x) = \sum_{k=0}^{m} t_k(x) = \sum_{k=0}^{m} \frac{f^{(k)}(a)}{k!}(x-a)^k.$$

T_m is known as a Taylor polynomial. It matches f and its first m derivatives at a, so it is also known as an mth-order approximation. The infinite expansion T_∞ is known as a Taylor series. A Taylor series around $a=0$ is known as a Maclaurin series.

Although the notation is ugly, it isn't hard to work with. To confirm the derivative-matching, note that

$$t_k^{(n)}(x) = \begin{cases} f^{(k)}(a)\dfrac{(x-a)^{k-n}}{(k-n)!} & \text{for } n \leq k \\ 0 & \text{for } n > k \end{cases}.$$

In particular, $t_k^{(n)}(a)$ equals $f^{(n)}(a)$ for $n=k$ and 0 otherwise. It follows that $T_m^{(n)}(a) = f^{(n)}(a)$ for all $n \leq m$.

Any f that matches T_∞ is called analytic. For example, $f(x) = 1/(1-x)$ is analytic for $|x| < 1$ as it matches $T_\infty(0) = \sum_{k=0}^{\infty} x^k$. Here Newton-Raphson iteration of $1 - x - z^{-1}$ from $z_0 = 1$ implies the same expansion. The seed matches T_0, $z_1 = 1 + x$ matches T_1, $z_2 = (1+x)(1+x^2)$ matches T_3, $z_3 = (1+x)(1+x^2)(1+x^4)$ matches T_7, and so on.

Mean Value Theorem

How big is the gap between f and T_m? The following equation helps quantify it. If $f^{(m+1)}$ is well-defined between a and x, some w in that interval satisfies

$$f(x) - T_m(x) = f^{(m+1)}(w)\frac{(x-a)^{m+1}}{(m+1)!}. \tag{FGT}$$

I will call this the First Gap Theorem, or FGT for short. For the simplest case $m = 0$ and $f(x) = f(a)$, FGT boils down to the proposition that some interior point must be stationary. This is Rolle's Theorem, which we met in Chapter 3.1. To prove it, note that f must reach an extreme value at some interior w. If $f'(w) \neq 0$, we could make f more extreme by moving slightly above or below w. Hence $f'(w) = 0$.

The next-simplest case, known as the Mean Value Theorem, retains $m = 0$ but allows $f(x) \neq f(a)$. It guarantees a w for which

$$f'(w) = \frac{f(x) - f(a)}{x - a}.$$

In other words, if f is differentiable between a and x, at least one slope will match the average slope s. To prove this, form the secant $f(a) + s(z - a)$ from $(a, f(a))$ to $(x, f(x))$. Then subtract the secant from f to make a new function g. Since $g(a) = g(x) = 0$, Rolle's Theorem guarantees a w with $g'(w) = 0$, which implies $f'(w) = s$.

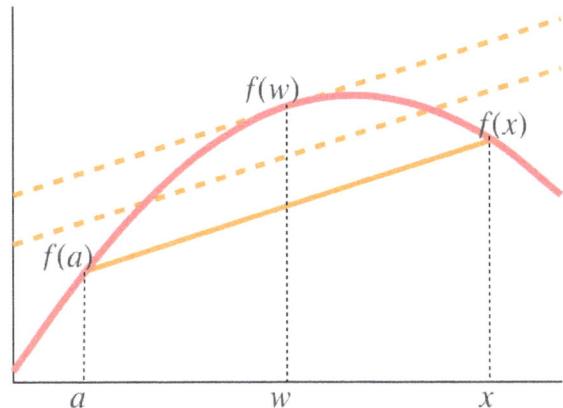

To find w graphically, draw lines parallel to the secant until one just touches f. The tangent point marks $(w, f(w))$.

Higher Precision

The next two paragraphs show how well higher-order Taylor polynomials can fit a function and why they sometimes don't. The gap between f and T_m can be formulated more precisely as

$$f(x) - T_m(x) = \int_a^x \frac{f^{(m+1)}(z)}{m!}(x - z)^m \, dz, \qquad \text{(SGT)}$$

which I call the Second Gap Theorem, or SGT for short. Define $Q_m(x) = \int_a^x \frac{(x-z)^m}{m!} dz = \frac{(x-a)^{m+1}}{(m+1)!}$ and $G = \frac{f(x) - T_m(x)}{Q_m(x)}$. SGT implies that G lies between the lowest and highest values of $f^{(m+1)}$ over the interval. Since $f^{(m+1)}$ is continuous, there must be some interior w for which $f^{(m+1)}(w) = G$. That is, if SGT is true, FGT must be true too.

CHAPTER 5

To prove SGT, we start with $m=0$, where SGT boils down to the Fundamental Theorem. Given any m for which SGT holds,

$$f(x) - T_{m+1}(x) = \int_a^x u\,dv - t_{m+1}(x),$$

where $u(z) = f^{(m+1)}(z)$ and $dv(z) = \left((x-z)^m/m!\right)dz$. Integrating by parts, $\int_a^x u\,dv = u\big|_a^x - \int_a^x v\,du$, where $v(z) = -(x-z)^{m+1}/(m+1)!$, $du = f^{(m+2)}(z)dz$, and $uv\big|_a^x = f^{(m+1)}(a)Q_m(x) = t_{m+1}(x)$. Substitution confirms that SGT holds for $m+1$. By induction, SGT holds for any whole number m.

Are your eyes glazing over? Don't worry. You just need to appreciate that

- Taylor polynomials T_m can approximate any smooth function.
- the fit close to a improves as m rises.
- the fit can fail if $|x-a|$ gets large or higher derivatives soar.

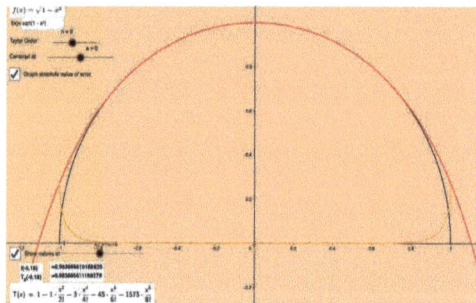

The GeoGebra activity "Taylor Polynomials" illustrates this.

Numerical Estimation

The quadratic T_2 approximation can help improve estimation of slopes and areas. Let's start with the $\Delta f/\Delta x$ estimate for f'. Substituting $\delta = x - a$, the error is

$$\frac{f(a+\delta) - f(a)}{\delta} - f'(a) \approx \tfrac{1}{2} f''(a)\delta.$$

If we form another estimate using a nudge $-\delta$ and average the two together, the terms in f'' offset each other, so that

$$f'(a) \approx \frac{f(a+\delta) - f(a-\delta)}{2\delta}.$$

This is called the central difference estimator for f'. It uses a secant centered across $(a, f(a))$ rather than a secant to or from $(a, f(a))$.

The central estimator can be much better than its constituents. For example, when $f(x) = x^3$ and we estimate $f'(4)$ using $\delta = 1$, the central estimate $\frac{1}{2}(f(5) - f(3)) = 49$ is much closer to the true value 48 than the right estimate $f(5) - f(4) = 61$ or the left estimate $f(4) - f(3) = 37$.

A comparable integral estimator is known as Simpson's rule. To derive it, let's start with a single Riemann slice stretching from $a - \delta$ to $a + \delta$. The midpoint rule treats the Riemann slice as a rectangle of width 2δ and height $f(a)$. The trapezoidal rule treats it as a trapezoid of average height $\frac{1}{2}(f(a-\delta) + f(a+\delta)) \approx f(a) + \frac{1}{2}f''(a)\delta^2$. Simpson's rule treats the slice as quadratic at the top. The underlying area matches the integral of the T_2 approximation to f, namely

$$\left(f(a)z + \tfrac{1}{2}f'(a)z^2 + \tfrac{1}{6}f''(a)z^3 \right)\Big|_{-\delta}^{\delta} = 2\delta\left(f(a) + \tfrac{1}{6}f''(a)\delta^2 \right).$$

Hence Simpson's Rule amounts to taking two-thirds of the midpoint rule plus one-third of the trapezoidal rule. To make this clearer, let's view each rule as taking a weighted average of $f(a-\delta)$, $f(a)$ and $f(a+\delta)$. The weights are $(0, 1, 0)$ for the midpoint rule, $(\frac{1}{2}, \frac{1}{2})$ for the trapezoidal rule, and $(\frac{1}{6}, \frac{4}{6}, \frac{1}{6})$ for Simpson's rule. When we join together all the Riemann slices, the relative weights become:

- midpoint rule: $(0, 1, 0, 1, \ldots, 0, 1, 0)$.
- trapezoidal rule: $(1, 2, 2, 2, \ldots, 2, 2, 1)$.
- Simpson's rule: $(1, 4, 2, 4, \ldots, 2, 4, 1)$.

Simpson's rule is remarkably effective. For example, if we use only $x = \{0, 1, 2, 3, 4\}$ to approximate $\int_0^4 x^4 dx$, it overestimates the true area by less than 0.3%, whereas the midpoint rule underestimates by 20% and the trapezoidal rule overestimates by 10%. Doubling the slices trims the Simpson error to under 0.02%.

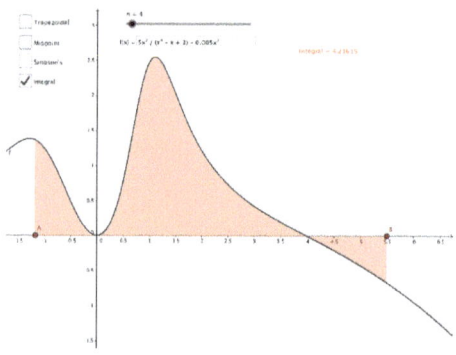

Compare midpoint, trapezoidal and Simpson's rule estimates in the GeoGebra activity "Numeric Integration".

5.3. CONVERGENCE

Ambrosia

Punch was wary when Judy entered the room, but she was all smiles. *"Don't worry, Punch, I'm not one to hold a needless grudge. And to prove it, I've brought a cake to share with you."*

"Really?" asked Punch. He leaned over to inspect it. *"Wow, that looks scrumptious, and the aroma is out of this world."*

"Ambrosia," said Judy, *"food of the gods. Nothing is better."* She handed him a cake knife. *"Here, cut yourself a big slice."*

Punch smacked his lips. *"Thanks, Judy. I wasn't expecting this. But don't you want to do the cutting? It's your cake."*

"Yes, it is, but I don't mind as long as you give me half of every slice you cut for yourself."

"That's more than fair," said Punch. He sliced the cake straight down the middle and gave half to Judy. She ate with such gusto that he stuck a fork into his half and ...

"Wait a second," said Judy. *"Isn't that a slice on your plate?"*

"Of course. I already gave you the other half."

"Fine, but that was half of the whole. I want half of your slice too, just like you promised."

"I get it," said Punch after a pause. "You want three-quarters in all, not just half. Why didn't you say that from the start?" He gave her half of his half. "I'm still happy with my quarter slice."

"I bet you are," said Judy. "Only a quarter slice is still a slice. Give me half of that too."

"But that will leave me with just an eighth, which you'll want half of too, and then half of the sixteenth that's left, and then… and then… Hey, there won't be anything left for me. You tricked me!"

"Maybe so," said Judy. "But here's a chance to get even. I brought a second ambrosia cake, just as good as the first. I'll still take a fixed positive share $1-r$ of every slice but you can choose any fractional r you want. Deal?"

Punch agreed, chose $r = 0.99$ and… came out empty-handed again. Every share that Judy took left him a fraction r of what he had before. After k slices his share of the original was r^k and was bound to shrink more. The limit couldn't be anything positive, because if it were, Judy would take some of that too.

Since Punch's share is 0, Judy's share is 1. Let's look at how it is composed. She receives a share $1-r$ from the whole, $(1-r)r$ from the first slice of r, $(1-r)r^2$ from the second slice of r^2 and so on. Hence

$$(1-r)1 + (1-r)r \\ + (1-r)r^2 + \cdots = 1.$$

To simplify, divide everything by the common factor $1-r$:

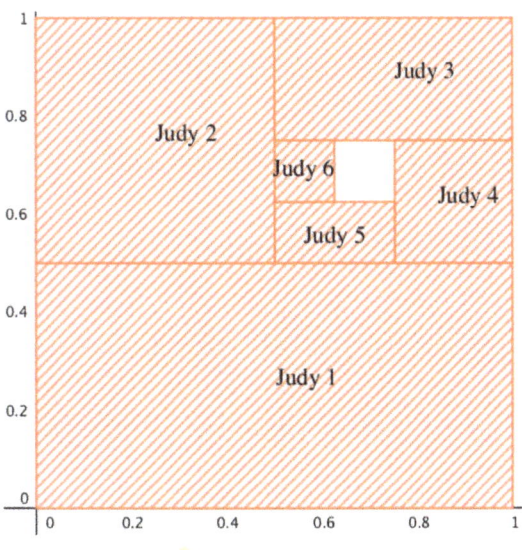

CHAPTER 5

$$1+r+r^2+\cdots = \sum_{k=0}^{\infty} r^k = \frac{1}{1-r}.$$

Now it is clear that an infinite series can sum to a finite number even when every term is positive. Multiplying every term by some seed s won't change that. The generalized version is known as a geometric series.

Yet this equation cannot always hold. For $r>1$, $(1-r)^{-1}$ is negative while the series grows without bound. For $r<0$, define $s=1+r$ and regroup into the geometric series $s+sr^2+sr^4+\cdots$. This has multiplier r^2, so it converges for $|r|<1$ and diverges for $|r|>1$. At $r=-1$ the base series alternates between 1 and 0.

Types of Convergence

The successive partial sums $S_n(r) = \sum_{k=0}^{n} sr^k$ form an infinite sequence $S(r) = \{S_0(r), S_1(r), S_2(r), ...\}$. For $s>0$ and $r>0$, $S(r)$ is monotone increasing, which means that each element exceeds the element before. For $s<0$ and $r>0$, $S(r)$ is monotone decreasing. For $r<0$, $S(r)$ alternates at every step between increases and decreases. These sequences can converge or fail to converge in eight distinct ways, which are best appreciated when graphed:

- For $r \geq 1$ and $s>0$,
  $S(r)$ rises without bound.

- For $r \geq 1$ and $s<0$,
  $S(r)$ falls without bound.

- For $0 < r < 1$ and $s > 0$, $S(r)$ converges from below to a limit.

- For $0 < r < 1$ and $s < 0$, $S(r)$ converges from above to a limit.

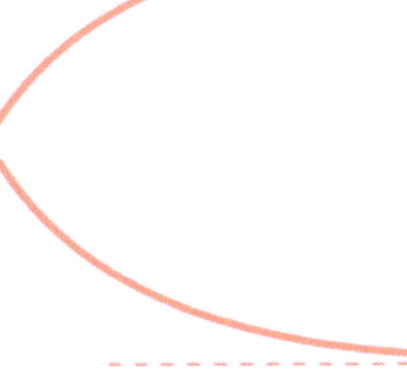

- $S(0)$ is constant.

- For $-1 < r < 0$, $S(r)$ oscillates in a shrinking band to a limit.

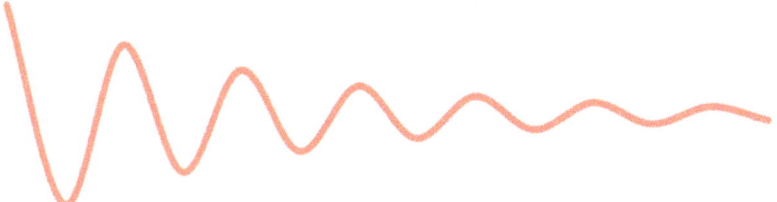

- For $r = -1$, $S(r)$ oscillates in a stable band.

- For $r \leq -1$, $S(r)$ oscillates in a stable or ever-widening band.

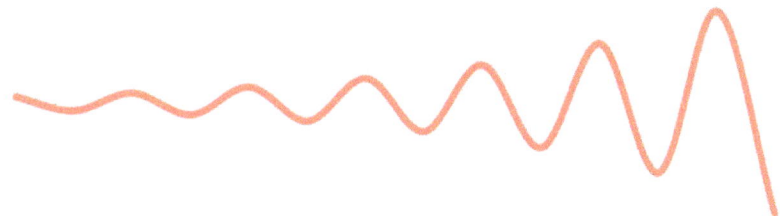

CHAPTER 5

Ratio Tests

Geometric sums offer guidance on convergence. Consider an arbitrary series $z_0 + z_1 + z_2 + \cdots$ and look at the ratios $r_k = z_k/z_{k-1}$. Suppose that $|r_k| \geq 1$ for every k beyond some threshold K. Then the $|z_k|$ will increase and the partial sums will never draw close enough to converge.

Another possibility is that $0 < r_k < 1 - \delta < 1$ for every $k > K$. Then the tail is bounded by a geometric series with seed z_K and ratio $1 - \delta$ that converges to z_K/δ. For $z_K > 0$, this sets an upper bound for the monotone increasing partial sums, which guarantees convergence. For $z_K < 0$, the same argument applies to the $-z_k$ and we just flip the sign at the end.

If the z_k never settle down to the same sign, combine like-signed neighbors to obtain a series that alternates signs at every step. Let us write this as $w_0 - w_1 + w_2 - w_3 + \cdots$ with all $w_k \geq 0$. If the w_k are monotone decreasing for $k \geq K$, then bump up K to an even number and express the tail sum in two different ways:

Upward: $(w_K - w_{K+1}) + (w_{K+2} - w_{K+3}) + (w_{K+4} - w_{K+5}) + \cdots$.

Downward: $w_K - (w_{K+1} - w_{K+2}) - (w_{K+3} - w_{K+4}) - \cdots$.

Since all the differences in parentheses are positive, *Upward* is monotone increasing while *Downward* is monotone decreasing. They squeeze toward a common limit if the w_k converge to zero. If $|w_k|$ never falls below some $\delta > 0$, the series will oscillate in a band at least δ wide.

If the r_k converge to 1 from below, the series may or may not converge. Here are two examples:

- $\dfrac{1}{1 \cdot 2} + \dfrac{1}{2 \cdot 3} + \dfrac{1}{3 \cdot 4} + \cdots = 1$. To confirm, note that $\dfrac{1}{k(k+1)} = \dfrac{1}{k} - \dfrac{1}{k+1}$.
  Substitution yields $\frac{1}{1} - \frac{1}{2} + \frac{1}{2} - \frac{1}{3} + \frac{1}{3} - \frac{1}{4} + \cdots = 1 + 0 + 0 + \cdots$.

- $\tfrac{1}{2} + \tfrac{1}{3} + \tfrac{1}{4} + \cdots = \infty$. To confirm, raise each denominator to the next higher power of 2, which yields $\tfrac{1}{2} + \tfrac{1}{4} + \tfrac{1}{4} + \tfrac{1}{8} + \tfrac{1}{8} + \tfrac{1}{8} + \tfrac{1}{8} + \cdots$. The two values of $\tfrac{1}{4}$ sum to $\tfrac{1}{2}$, the four values of $\tfrac{1}{8}$ sum to $\tfrac{1}{2}$, and so on. Hence the original sum exceeds $\tfrac{1}{2} + \tfrac{1}{2} + \tfrac{1}{2} + \cdots = \infty$.

For Taylor series, the relevant $z_k = f^{(k)}(a)(x-a)^k/k!$ imply ratios of $r_k = \dfrac{(x-a)f^{(k)}(a)}{k f^{(k-1)}(a)}$. If $q_k = \dfrac{f^{(k)}(a)}{k f^{(k-1)}(a)}$ rises or falls without bound as k increases, the Taylor series will fail to converge even for tiny $x-a$. However, that is rare. Usually the q_k converge in absolute value to some constant Q. If so, the Taylor series will converge for any $|x-a| < 1/Q$. This bound is known as the radius of convergence around a. Here are several Taylor series that converge inside radius 1:

- In the expansion $\dfrac{1}{1-x} = \sum_{k=0}^{\infty} x^k$, $q_k = 1$ for all k.

- In the expansion $\dfrac{1}{(1-x)^2} = \sum_{k=0}^{\infty} (k+1) x^k$, $q_k = 1 + \dfrac{1}{k}$.

- In the expansion $\sqrt{1+x} = 1 + \dfrac{x}{2} - \dfrac{x^2}{8} + \dfrac{x^3}{16} - \cdots$, $q_k = \dfrac{3}{2k} - 1$.

For a break from symbols, let's review some useful vocabulary.

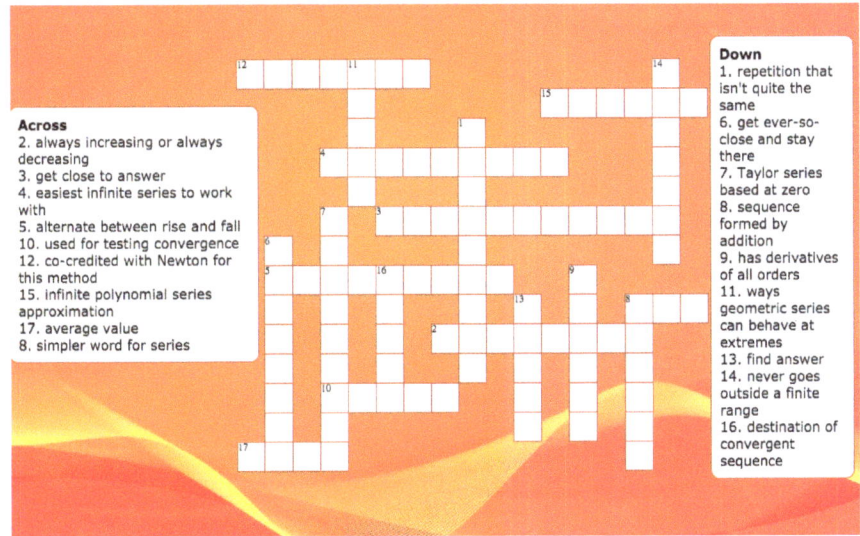

Across
2. always increasing or always decreasing
3. get close to answer
4. easiest infinite series to work with
5. alternate between rise and fall
10. used for testing convergence
12. co-credited with Newton for this method
15. infinite polynomial series approximation
17. average value
8. simpler word for series

Down
1. repetition that isn't quite the same
6. get ever-so-close and stay there
7. Taylor series based at zero
8. sequence formed by addition
9. has derivatives of all orders
11. ways geometric series can behave at extremes
13. find answer
14. never goes outside a finite range
16. destination of convergent sequence

Chapter 5

Binomial Theorem

How many ways can we choose k balls from a set of n distinct balls? We have n choices for the first ball, $n-1$ choices for the second, and so on down to $n-k+1$ choices for the kth ball. That makes $n(n-1)\cdots(n-k+1)$ distinct orderings, called permutations. If we care only about which balls are chosen, we need to divide by the $k!$ ways that k distinct objects can be ordered. The quotient indicates the number of distinct combinations of k objects out of n and is written:

$$\binom{n}{k} = \frac{n(n-1)\cdots(n-k+1)}{k(k-1)\cdots 1}.$$

The values can be displayed in a pyramid that makes them easy to calculate. It is often called Pascal's triangle, although it was discovered and rediscovered many times before Pascal. While each value simply adds the two entries above it, the $k+1^{st}$ entry on row $n+1$ is $\binom{n}{k}$.

Here is one way to prove the connection. Compared to $\binom{n}{k-1}$, $\binom{n}{k}$ has extra factors of $n-k+1$ in the numerator and k in the denominator. Hence their sum is $\binom{n}{k-1} \cdot \left(\frac{n-k+1}{k}+1\right) = \binom{n}{k-1} \cdot \frac{n+1}{k} = \binom{n+1}{k}$.

These numbers are coefficients in the binomial expansion $(w+x)^n = \sum_{k=0}^{n} \binom{n}{k} w^{n-k} x^k$. Remarkably, an infinite extension of this formula holds for all n, integer or not, provided $|x|<|w|$. All we need to do is let k keep increasing past n:

$$(w+x)^n = \sum_{k=0}^{\infty} \binom{n}{k} w^{n-k} x^k.$$

First discovered by Newton, the generalized binomial theorem amounts to a Maclaurin series expansion, as sketched in these calculations.

$$f(z) = (1+z)^n, \quad f'(z) = n(1+z)^{n-1}$$
$$f''(z) = n(n-1)(1+z)^{n-1}$$
$$f^{(k)}(z) = n(n-1)\cdots(n-k+1)(1+z)^{n-k}$$
$$f(z) = \sum_{k=0}^{\infty} \frac{f^{(k)}(0)}{k!} z^k = \sum_{k=0}^{\infty} \binom{n}{k} z^k$$
$$(w+x)^n = w^n (1+x/w)^n = w^n f(x/w)$$
$$= \sum_{k=0}^{\infty} \binom{n}{k} w^{n-k} x^k$$

5.4. TAYLOR CALCULUS

Constant Relative Growth

So far we have used Taylor series as approximations. However, Taylor series can also help us find exact solutions to calculus problems. For example, suppose the rate of change is always a constant share b of the current value. In symbols, $f' = bf$. What can we say about f?

Observation 1: f can't be a simple power function. If $f(x) = cx^n$, the relative growth rate $f'/f = n/x$ shrinks toward 0 as x grows.

Observation 2: f can't be a finite polynomial, since the latter's growth is ultimately driven by the highest power of x.

Observation 3: Differentiating both sides of $f' = bf$ shows that $f'' = bf' = b^2 f$. By induction, $f^{(k)} = b^k f$ for every positive integer k.

Observation 4: After substituting for $f^{(k)}$, the Maclaurin series for f equals some constant $A = f(0)$ times $\sum_{k=0}^{\infty} \frac{b^k}{k!} x^k = 1 + bx + \frac{b^2 x^2}{2} + \frac{b^3 x^3}{6} + \cdots$.

Observation 5: The ratios of neighboring terms are $r_n = bx/n$. Since $|r_n| < 1$ for every $n > |bx|$, the Maclaurin series for f converges for all x.

Observation 6: Defining $g(x) = \sum_{k=0}^{\infty} \frac{x^k}{k!} = 1 + x + \frac{x^2}{2} + \frac{x^3}{6} + \cdots$, every f takes the form $f(x) = A g(bx)$.

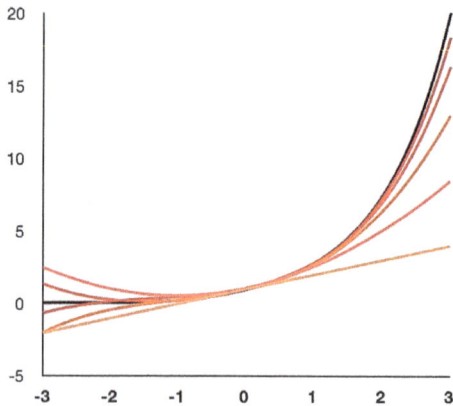

The series $\sum_{k=0}^{\infty} \frac{x^k}{k!}$ is so crucial to higher math that it has a special name: exponential. It can be written $\exp(x)$. It is charted here in black along with the corresponding Taylor polynomials from order 1 to order 5.

For $x > 0$, convergence looks smooth. Each higher power k of x tilts the curve upward, with the added term having x/k times the impact of the term before. For $x < 0$, convergence looks messy. While $\exp(x)$ shrinks monotonically to 0 as we move left, every finite polynomial veers off to $+\infty$ or $-\infty$. For neater convergence with $x < 0$, apply the identity $\exp(-x) = 1/\exp(x)$. To prove that identity, note that

$$\frac{d}{dx}\left(\exp(x) \cdot \exp(-x)\right) = \exp(x) \cdot \exp(-x) - \exp(x) \cdot \left(-\exp(-x)\right) = 0.$$

Hence $\exp(x) \cdot \exp(-x) = \exp(0) \cdot \exp(-0) = 1$ for all x.

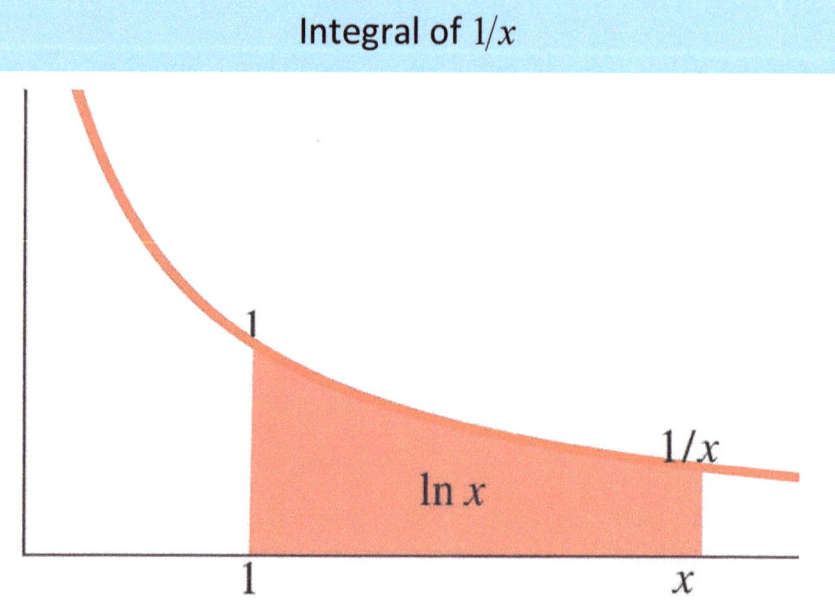

Integral of $1/x$

Another function crucial to higher math is the natural logarithm

$$\ln x = \ln(x) = \int_1^x \frac{dz}{z},$$

which is well-defined for any $x > 0$. While we can compute it to high precision using Riemann sums, we would like to find crisper expressions. Unfortunately, the formula $\int_1^x z^{p-1} dz = (x^p - 1)/p$ works for every p except the $p = 0$ we need.

Let's define $z = x - 1$ and apply the Maclaurin series

$$\frac{1}{1+z} = 1 - z + z^2 - z^3 + \cdots.$$

Integrating term by term,

$$\ln x = \ln(1 + z) = z - \frac{z^2}{2} + \frac{z^3}{3} - \cdots.$$

This converges for $-1 < z \leq 1$, which corresponds to $0 < x \leq 2$. For larger x, note that $dx^{-1}/x^{-1} = -x^{-2}dx/x^{-1} = -dx/x$. Hence $\ln x = -\ln x^{-1}$. If we define $w = 1 - x^{-1}$, substituting $-w$ for z in the previous series shows that

$$\ln x = -\ln(1 - w) = w + \frac{w^2}{2} + \frac{w^3}{3} + \cdots.$$

This converges for $-1 \leq w < 1$ or $x \geq \frac{1}{2}$. A faster-converging series that works for any x, and which provides a good second-order approximation with a single term, is

$$\ln x = 2v + \frac{2v^3}{3} + \frac{2v^5}{5} + \cdots \text{ for } v = \frac{x-1}{x+1}.$$

Here are confirming calculations. They use one property that we have not yet proved, namely that $\ln(a/b) = \ln a - \ln b$. Don't bother to memorize the results. Do marvel at how Taylor series can make hard calculations easier.

$$v = \frac{x-1}{x+1} \rightarrow x = \frac{1+v}{1-v}$$
$$\ln(1+v) = v - \tfrac{1}{2}v^2 + \tfrac{1}{3}v^3 - \cdots$$
$$\ln(1-v) = -v - \tfrac{1}{2}v^2 - \tfrac{1}{3}v^3 - \cdots$$
$$\ln x = \ln(1+v) - \ln(1-v)$$
$$= 2v + \tfrac{2}{3}v^3 + \tfrac{2}{5}v^5 + \cdots$$

CHAPTER 5

Springs

A mechanical spring stores energy when compressed or stretched and releases it when the pressure is removed. The physicist Robert Hooke noted that a spring tends to extend or contract in proportion to the force applied to it. Standard terminology for Hooke's Law is $F = -kx$, where F denotes the force, k the stiffness of the spring, and x the extension from resting position.

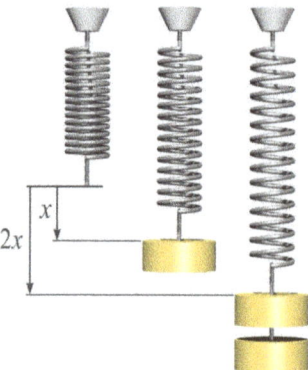

If we hang a disk on a vertical spring and lower it gently, the spring will extend by $1/k$ times the disk weight mg. A disk resting there will be stable because retraction from the spring exactly offsets gravity. A disk that drops from higher will overshoot the resting point. The spring's retraction force will build until it slows the disk to a halt and pulls it back up.

To focus on string behavior without gravity, let's turn the spring horizontally and replace the weights with a frictionless box. By Newton's Second Law, F equals box mass m times the acceleration $x''(t)$. Combining that with Hooke's Law, $x''(t) = -cx(t)$ for $c = k/m$.

Given an initial position $x_0 = x(0)$ and an initial speed $v_0 = x'(0)$, the equation above determines $x(t)$ for every t. But what kind of function is $x(t)$? It can't be polynomial, exponential or logarithmic, since none of those keep oscillating. We need something that resembles a wave.

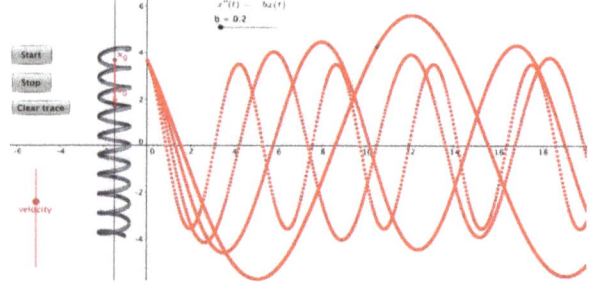

Make waves with the GeoGebra activity "Oscillating Spring". How do x_0, v_0 and c affect the waves?

Computing the Taylor series is easy. Differentiate both sides of $x'' = -cx$ repeatedly to show that $x^{(n+2)}(t) = -cx^{(n)}(t)$ for all n. This generates one chain of equations for even derivatives and another chain for odd derivatives. The even chain implies $x^{(2n)}(t) = (-1)^n c^n x(t)$. The odd chain implies $x^{(2n+1)}(t) = (-1)^n c^n x'(t)$. The resulting Maclaurin expansion is

$$x(t) = x_0 + v_0 t - x_0 \frac{ct^2}{2!} - v_0 \frac{ct^3}{3!} + x_0 \frac{c^2 t^4}{4!} + \cdots.$$

To tidy this up, it helps to consider two simpler cases. The first, called cosine, sets $v_0 = 0$ and $x_0 = c = 1$:

$$\cos t = \cos(t) = \sum_{k=0}^{\infty} \frac{(-1)^k t^{2k}}{(2k)!} = 1 - \frac{t^2}{2!} + \frac{t^4}{4!} - \cdots.$$

The second case, called sine, sets $x_0 = 0$ and $v_0 = c = 1$:

$$\sin t = \sin(t) = \sum_{k=0}^{\infty} \frac{(-1)^k t^{2k+1}}{(2k+1)!} = t - \frac{t^3}{3!} + \frac{t^5}{5!} - \cdots.$$

For either sine or cosine, the ratio of neighboring Taylor terms is approximately $-t^2/(4k^2)$, which exceeds -1 for $2k > |t|$ and guarantees convergence. Differentiate to see that $(\sin t)' = \cos t$ and $(\cos t)' = -\sin t$.

The Maclaurin series for the spring can then be rewritten as

$$x(t) = x_0 \cos \sqrt{c} t + \left(v_0 / \sqrt{c}\right) \sin \sqrt{c} t.$$

To simplify, let t_1 denote a time when extension peaks. Since $v(t_1) = 0$, $x(t) = x(t_1) \cos \sqrt{c}(t - t_1)$. Alternatively, given time t_2 when $x(t_2) = 0$, $x(t) = v(t_2) \sin \sqrt{c}(t - t_2)$. Hence sines must be shifted versions of cosines, although we don't yet know the appropriate shift.

Here is a chart of $\cos t$ and its Taylor expansions out to t^{14}.

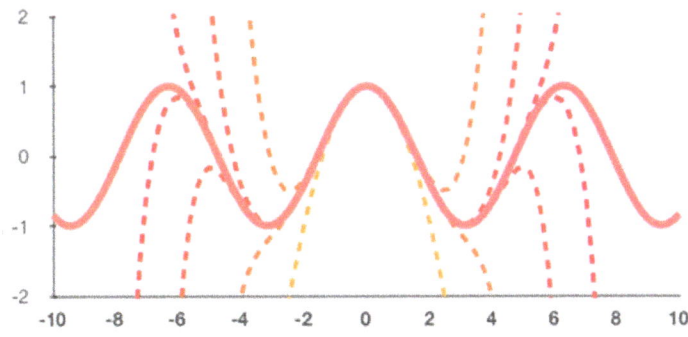

CHAPTER 5

Term by Term

Differential equations contain functions and their derivatives. Taylor series can help us solve them even when we can't work out a crisp formula for all the terms. Here's how it works:

Step 1: Replace every function $f(x)$ with a Taylor series $\sum_{k=0}^{\infty} \frac{c_k}{k!} x^k$.

Step 2: Replace every derivative $f^{(n)}(x)$ with the corresponding Taylor series $\sum_{k=0}^{\infty} \frac{c_{k+n}}{k!} x^k$.

Step 3: Combine like terms to rewrite the differential equation in the form " $G(x) = \sum_{k=0}^{\infty} g_k(c_1, c_2, ...) x^k = 0$ for all x ".

Step 4: Choose the not-yet-determined c_j to equate each g_k to zero.

Step 5: Discard candidate solutions that violate other requirements.

To show that step 3 implies step 4, note that $G^{(k)}(0) = g_k k! = 0$ for all k. The same logic explains why any Taylor series f and h that match over even a short range must have identical coefficients, since $f - h$ must have coefficients of zero.

For an application, suppose a spring experiences a frictional force proportional to velocity. Its equation of motion now takes the form $x''(t) + ax'(t) + bx(t) = 0$. If $x(t) = \sum_{k=0}^{\infty} (c_k/k!) t^k$, then $c_{k+2} + ac_{k+1} + bc_k = 0$ for every k. Given initial position and initial velocity we can solve this recursively for every c_k. Chapter 6.4 will derive a neater solution. However, if we just want to compute motion the Taylor series will suffice.

Only one Taylor series can generate exactly this shape.

L'Hôpital's Rule

While limits of ratios help determine Taylor series, Taylor series often help us determine the limits of ratios. Consider for example

$$\lim_{x \to 0} \frac{\sqrt{1+x}-1}{x}.$$

Using the T_1 approximation $\sqrt{1+x} \approx 1 + \tfrac{1}{2}x$, the ratio simplifies to $\tfrac{1}{2}$. In general,

$$f(a) = g(a) = 0 \text{ implies } \lim_{x \to a} \frac{f(x)}{g(x)} = \frac{f'(a)}{g'(a)},$$

provided the latter is well-defined. This is known as L'Hôpital's rule. Limits of Taylor series provide an intuitively appealing proof:

$$\frac{f(a+\delta)}{g(a+\delta)} \approx \frac{0 + f'(a)\delta}{0 + g'(a)\delta} = \frac{f'(a)}{g'(a)}.$$

Although the proof is too complicated to present here, L'Hôpital's rule also applies for $f(a) = g(a) = \infty$. For example, $\lim_{x \to \infty} \frac{\exp(x)}{x} = \frac{\exp(\infty)}{1} = \infty$. By induction, $\lim_{x \to \infty} \frac{\exp(x)}{x^m} = \lim_{x \to \infty} \frac{\exp(x)}{mx^{m-1}} = \infty$ for any positive integer m. Hence $\exp(x)$ grows faster than any finite polynomial.

$$\frac{\exp(x)-1}{x} \to \frac{\exp(0)}{1} = 1$$

$$\frac{\ln(1+x)}{x} \to \frac{1/(1+0)}{1} = 1$$

$$\frac{\sin x}{x} \to \frac{\cos 0}{1} = 1$$

Here are more applications of L'Hôpital's rule as x approaches 0.

$$\frac{\ln x}{x} \to \frac{1/\infty}{1} = 0$$

Here is an application of L'Hôpital's rule as x approaches ∞.

Sometimes we need to apply L'Hôpital's rule several times in a row. This amounts to higher-order Taylor approximations.

$$\frac{\exp(x) - x - 1}{x^2} \to \frac{\exp(x) - 1}{2x} \to \frac{\exp(0)}{2} = \tfrac{1}{2}$$

$$\frac{1 - \cos x}{x^2} \to \frac{\sin x}{2x} \to \frac{\cos 0}{2} = \tfrac{1}{2}$$

6

Logs and Antilogs

Logarithms convert products into sums and ratios into differences. Their exponential inverses model constant proportional growth. Calculus makes logs and antilogs easier to understand and to apply to practical problems.

6.1. LOGARITHMS

Wonderful

One day Punch and Judy went hiking and got so busy arguing that they wandered into a wormhole in time. Emerging a century before Newton, Punch spotted opportunity and landed a Senior Arithmetician job in the kingdom's Budget Department. Before long he was miserable.

"Long multiplication is so annoying," he told Judy. *"Do you know how easy it is to shift the wrong number of decimal places or mix up 9×6 with 8×7? Long division is even worse, as each step needs trial multiplications. I miss calculators!"*

"Dry up, Punch, they won't be invented for centuries. How about wishing for something simpler, like functions that convert multiplication and division into addition and subtraction?"

"Do you mean an f for which $f(xy) = f(x) + f(y)$ for any x and y?"

"Yes," said Judy, "if they're positive. Read off $f(x)$ and $f(y)$ from a table of f values, add them, find the nearest value in the table and convert back to xy. Napier said he's been building the table for a year."

"Wonderful! How soon will it be ready?"

"If I recall future history correctly, 19 more years."

"19? I wish he'd hurry up, and also create a second table to convert powers and roots into multiplication and division."

"Napier's f does that too. For any n, $f(x^n) = nf(x)$. In fact, that's how he's building it. He picked a tiny ε and defined $f(1+\varepsilon) = A$. His table just lists various $\langle (1+\varepsilon)^n, nA \rangle$ pairs. Sounds easy, except that he is computing millions of pairs."

"Genius: But which value of A is best?"

"For multiplication, it doesn't matter," said Judy. "Posterity will call $f(x)$ a logarithm to base b, written $\log_b x$ or $\log_b(x)$, if $f(b) = 1$. A popular choice will be $b = 10$. Mathematicians will prefer e."

"What's the point of e?"

"It makes $f'(1)$ equal 1. That logarithm will be called 'natural' and often written $\ln$ in tribute to the French version 'logarithme naturel'."

"How do we get from a table of $f(x)$ values to precise derivatives?"

We don't. Go back to $f(xy) = f(x) + f(y)$, hold x fixed, and differentiate both sides with respect to y to obtain $xf'(xy) = f'(y)$. Setting $y = 1$ shows that $f'(x) = c/x$ for $c = f'(1)$.

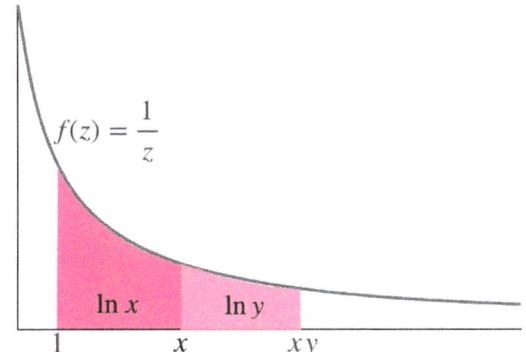

Since this holds for any x,

$$f(x) = c \int_1^x \frac{dz}{z}.$$

The integral defines $\ln x$."

CHAPTER 6

$\log x = \log(x \cdot 1) = \log x + \log 1 \to \log 1 = 0$

$\log 1 = \log(xx^{-1}) = \log x + \log x^{-1} \to \log x^{-1} = -\log x$

$\log x^2 = \log(x \cdot x) = \log x + \log x = 2 \log x$

$\log x = \log(x^{1/2})^2 = 2 \log x^{1/2} \to \log x^{1/2} = \tfrac{1}{2} \log x$

Here are examples demonstrating for various r that

$\log_b x^r = r \log_b x$.

The identity is readily proved using induction as described in Chapter 3.4. Better yet, define $y = z^r$, note that $dy = rz^{r-1}dz$ and compute

$$\ln x^r = \int_0^{x^r} \frac{dy}{y} = \int_0^x \frac{rz^{r-1}dz}{z^r} = r\int_0^x \frac{dz}{z} = r\ln x.$$

Slide Rules

In a linear scale, the distance between x and $x+c$ is the same for every x. In a logarithmic ("log") scale, the distance between x and cx is the same for every x.

A log scale makes 8 as far from 4 as 4 is from 2 or 2 is from 1. It places 10 midway between 1 and 100 and places $\sqrt{10} \approx 3.16$ midway between 1 and 10. Every x and y are separated in direct proportion to $\log y - \log x = \log(y/x)$.

Sliding identical log scales past each other can help us multiply two numbers, as in $2 \times 3 = 6$. A slide rule fastens log scales together to make them easy to use.

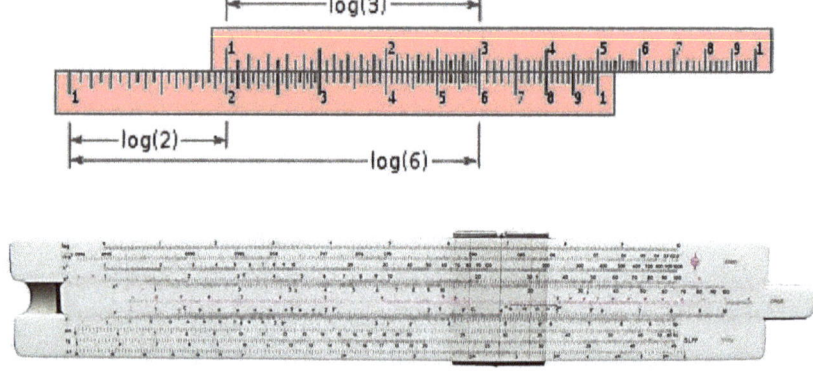

Longer slide rules offer more precision but can be unwieldy. The slide rule to the right spirals a 15-meter scale into a diameter of 20 centimeters. High-quality slide rules were prized in engineering. Not until the 1970s did electronic calculators get cheap enough to replace them.

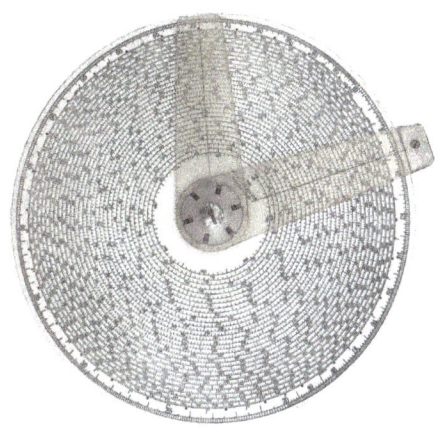

Loudness

Log scales are useful whenever the compared measurements vary by several orders of magnitude. For example, the sounds we hear every day sometimes differ in intensity by a factor of a billion or more. When compared in absolute magnitude, the loudest sounds make everything else seem infinitesimal.

The preferred measure is logarithmic. Decibels (dB) compare sound intensity I to a benchmark I_0, take the base-10 logarithm of the ratio and multiply by 10. That is, the decibels indicate $10 \log_{10}(I/I_0)$.

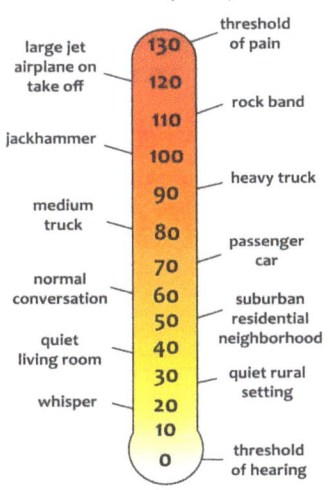

The benchmark 0 represents sound that most people can barely hear. Double the intensity and the decibels (dB) increase by three. However, humans don't hear all frequencies equally. Decibels adjusted for human sensitivity are known as dBA. Normal conversation is 60 dBA, at least 3000 times louder than a purr. Ratchet up loudness 1000-fold to 90 dBA, listen to it eight hours a day and risk long-term hearing damage. At 130 dBA your ears will scream in anguish.

CHAPTER 6

Earthquakes

In 1935, the seismologist Charles Richter was charting the absolute magnitudes M of California earthquakes and found the range unmanageably large. A colleague suggested plotting log magnitude instead. What became known as the Richter scale measured $\log_{10}(M/M_0)$, with M_0 so low that even a value of 3 was barely detectable.

An estimated 130,000 Richter 3 earthquakes occur every year. They are strong enough to make people feel the ground shake but they rarely cause damage. Each extra Richter unit multiplies M by ten and makes the quakes about ten times rarer. About 10 to 20 earthquakes per year are Richter 7 or higher. These are major earthquakes, bound to cause deaths and severe building damage in populated areas.

The tenfold differential between Richter units is misleading. The M refers to the amplitude of oscillation, while the released energy scales with $M^{3/2}$. As a result, Richter jumps of 2.0 signify a thousand times more power, not a hundred. A Richter 8 quake has an energy equivalent of 15 megatons of TNT, a million times more than Richter 4 and about a thousand times as powerful as the atom bomb dropped on Hiroshima. Fortunately for us, most of that energy is released underground.

Acidity

Ions are molecules with unequal numbers of positively charged protons and negatively charged electrons. Water solutions have measurable concentrations of each, particularly hydrogen ions H^+ or H_3O^+ and hydroxide OH^-. When H^+ dominates OH^-, solutions are called acidic. When OH^- dominates H^+, solutions are called alkaline.

The pH of a solution measures $-\log_{10}[H^+]$, where $[H^+]$ is the H^+ concentration in moles per liter. In pure water, H^+ and OH^- are equally concentrated at 10^{-7} moles per liter, so $pH = 7$ is considered neutral.

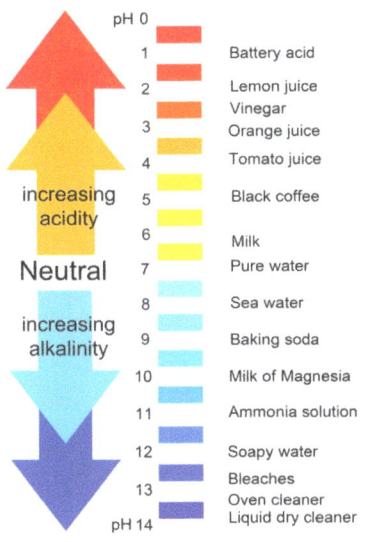

As solutions get more acidic, pH drops. Compared to pure water, $[H^+]$ is one thousand times higher in tomato juice at $pH = 4$ and one million times higher in battery acid at. Conversely, pH rises with alkalinity. Compared to pure water, $[H^+]$ is one thousand times lower in milk of magnesia at $pH = 10$ a million times lower in bleach at $pH = 13$.

Light

When nature spans many orders of magnitude but there's no consensus scale, we often describe it in powers of 10. This is equivalent to a $\log_{10}$ scale. A good example is light: not just visible light but all electromagnetic radiation. While all light acts in some respects like waves, its wavelength—the distance between neighboring crests and troughs—varies by up to a factor of a trillion trillion.

Visible light spans a narrow range of wavelengths: 400 to 700 nanometers (billionths of a meter). As wavelengths decrease, blue gives way to ultraviolet waves, which can be as short as a few nanometers. Then come X-rays, with waves as short as 0.01 nanometers. Gamma rays, generated by tremendous cosmic forces, sometimes squeeze ten million waves into a nanometer.

At longer wavelengths, red gives way to infrared waves, which can up to a millimeter long. Then come microwaves, which can be up to a meter long. FM radio waves are a few meters long. AM radio waves are about 100 meters long. Interstellar radio waves can be over 100,000 kilometers long.

CHAPTER 6

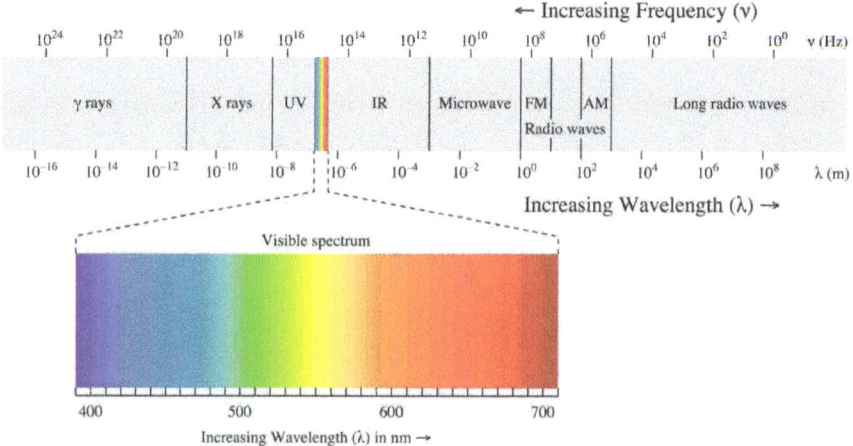

The frequency of electromagnetic waves equals their common speed of 300,000 kilometers per second divided by wavelength. Hence frequency spans just as broad a range as wavelength except in reverse order.

Relative Growth Rates

The Taylor series expansions of logarithms in Chapter 5.4 showed that $\ln x \to -\infty$ as $x \to 0$ and $\ln x = -\ln x^{-1} \to \infty$ as $x \to \infty$. We can write this informally as $\ln 0 = -\infty$ and $\ln \infty = \infty$. An alternative proof notes that

$$\int_0^1 \frac{dx}{x} = \int_\infty^1 \frac{dz^{-1}}{z^{-1}} = \int_\infty^1 \frac{-dz}{z^2 z^{-1}} = \int_1^\infty \frac{dz}{z} > \tfrac{1}{2} + \tfrac{1}{3} + \cdots = \infty.$$

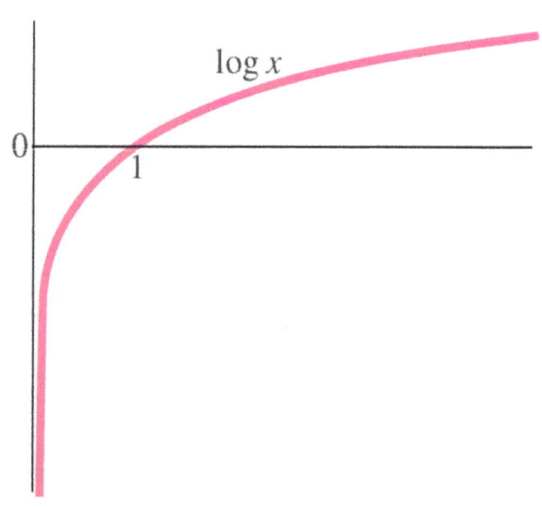

The slope $1/x$ indicates that $\ln x$ flattens from nearly vertical at tiny x to nearly horizontal at huge x. Since $\log_b x$ for any b equals $\ln x / \ln b$, all logarithms take the shape to the right.

For more insight into calculation, let's look at the difference between two logarithms. Often called log difference, it equals the logarithm of a ratio: $\ln z - \ln y = \ln(z/y)$. A first-order Taylor expansion indicates that

$$\ln \frac{z}{y} = \ln\left(1 + \frac{z-y}{y}\right) \approx \frac{z-y}{y}.$$

That is, log difference approximately equals relative or percentage change. For changes of less than 10%, the two measures are close. For example, $\ln 1.05 \approx 0.049$ and $\ln 0.95 \approx -0.051$.

While percentage change is easier to calculate, log difference has some advantages. Its magnitude doesn't depend on which value we choose as base, since $\ln(z/y) = -\ln(y/z)$. Also, the log changes over individual periods sum to the log change over the total period, since

$$\ln \frac{y_1}{y_0} + \cdots + \ln \frac{y_T}{y_{T-1}} = \ln \frac{y_T}{y_0}.$$

In contrast, neither property holds for percentage changes. For example, the percentage change to 8 from 4 equals $(8-4)/4 = +100\%$ while the percentage change back to 4 is $(4-8)/8 = -50\%$, so the two changes sum to $100\% - 50\% = 50\%$ rather than the 0% we want.

Chapter 5.4 noted that $\ln x$ is better approximated by $2(x-1)/(x+1)$ than by $w = x-1$. The ratio has Taylor expansion $w - \tfrac{1}{2}w^2 + \tfrac{1}{4}w^3 - \cdots$, which is accurate to second-order and nearly accurate to third-order (w^3 should be divided by 3 instead of by 4). Applying that to the log difference indicates that

$$\ln \frac{z}{y} \approx \frac{z-y}{\tfrac{1}{2}(y+z)}. \qquad \text{(APPROX)}$$

This compares the difference $z-y$ to the average of y and z rather than to the initial y alone. That can't be fully correct since it restricts $\ln x$ to values between -2 and $+2$. Still, APPROX is accurate to two digits for percentage changes of -33% to $+50\%$. Also, it preserves the desirable logarithmic property that aggregate change from y to z and back is zero.

6.2. ANTILOGARITHMS

Estimation of e

Suppose $b^y = x$ for some positive b. Taking the logarithm of both sides,

$$\log_b x = \log_b b^y = y \cdot \log_b b = y$$

Hence the inverse logarithm x, also called antilogarithm, is found by raising the logarithm y to its base b. For the natural logarithm, $e^{\ln x} = x$ and $\ln e^x = x$.

The base e of the natural logarithm is a magic number like π and at least as important. Here is one way to estimate it. For any integer n, APPROX indicates that

$$\frac{1}{n} = \ln e^{1/n} = \ln \frac{e^{1/n}}{1} \approx \frac{e^{1/n} - 1}{\frac{1}{2}(1 + e^{1/n})}.$$

Solving, $e^{1/n} \approx 1 + 1/(n - \frac{1}{2})$. Raise both sides to the nth power (if $n = 2^m$, just square the result m times) to obtain

$$e \approx \left(1 + \frac{1}{n - \frac{1}{2}}\right)^n.$$

This converges to 2.71828...

n	1	2	4	8	16	32	64
$e \approx$	3.0000	2.7778	2.7326	2.7218	2.7192	2.7185	2.7183

For large n we can drop the $-\frac{1}{2}$ and write

$$e = \lim_{n \to \infty} \left(1 + \frac{1}{n}\right)^n.$$

We can derive this more simply using the approximation $1/n = \ln e^{1/n} \approx e^{1/n} - 1$. While the expression doesn't converge nearly as quickly as the previous one, it has an appealing interpretation, which we will consider next.

Continuous Compounding

Imagine a bank offers the following contract: for every dollar we deposit for a 20-year period, it will repay two dollars at the end or maturity. One of those dollars covers the initial deposit or principal. The other is called interest. Its ratio per period to principal is called the rate of interest: in this case 100%.

Next imagine that a rival bank offers the same rate of interest but sweetens the deal by offering to pay half the interest every half-period. If we leave the first interest payment in the bank, our $1 + \frac{1}{2}(100\%) = \$1.50$ after 10 years will return $(1.50)^2 = \$2.25$ after 20 years. The difference is the \$0.25 interest paid on the first \$0.50 interest. This is known as "compound interest", in contrast to "simple interest" that earns no interest.

Next imagine that other banks offer the same rate of interest but sweeten the deal even more. Specifically, bank n offers to pay $100\%/n$ interest at n equally spaced intervals. Left untouched in the bank for the full 20 years, every dollar returns $(1+1/n)^n$ dollars at the end. In the limit of infinite n, the best deal repays e times the initial deposit.

Here is a related question. For every dollar in the account, ignoring rounding errors, how much money M is in the account one year later?

Since the rate of return stays constant for 20 years, $M^{20} = e$ or $M = e^{0.05}$. While 5% is the simple interest rate per year, the compounded yearly interest is $M - 1 \approx 5.127\%$. More generally, any dollar compounding continuously at rate of interest x returns e^x per period. The gross return over t periods is $(e^x)^t = e^{xt}$.

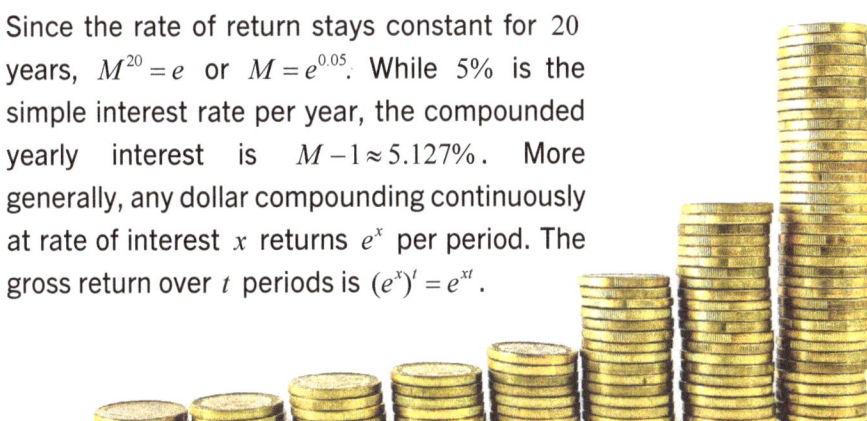

CHAPTER 6

Exponential Connections

One way to think of e^x is that it represents a principal of 1 plus countless layers of interest:

- Layer 1 is simple interest at rate x on the principal, with xt accumulated by time t.

- Layer 2 is simple interest at rate x on layer 1. It grows at rate $x \cdot xt$. The accumulation at time t is $\int_0^t x^2 t \, dt = \frac{x^2 t^2}{2}$.

- Layer 3 is simple interest at rate x on layer 2. It grows at rate $x \cdot \frac{x^2 t^2}{2}$. The accumulation at time t is $\int_0^t \frac{x^3 t^2}{2} dt = \frac{x^3 t^3}{3 \cdot 2}$.

- Layer n is simple interest at rate x on layer $n-1$. It grows at rate $x \cdot \frac{x^{n-1} t^{n-1}}{(n-1)!}$. The accumulation at time t is $\int_0^t \frac{x^n t^{n-1}}{(n-1)!} dt = \frac{x^n t^n}{n!}$.

- Hence the total return per period is $e^x = 1 + x + \frac{x^2}{2!} + \frac{x^3}{3!} + \cdots$. Notice anything familiar? You should. It's the exponential. Yes, $e^x = \exp(x)$.

Substituting $x = 1$ into the Taylor series, $e = 1 + 1 + \frac{1}{2!} + \frac{1}{3!} + \frac{1}{4!} + \cdots$, which converges much faster than the power functions used before:

Order	1	2	3	4	5	6	7
$e \approx$	2.000	2.5000	2.6667	2.7083	2.7167	2.7181	2.7183

The Taylor series also tells us that $(e^x)' = e^x$. However, we don't need Taylor series to prove this. Differentiate $x = \ln e^x$ using the chain rule to see that $1 = \frac{1}{e^x} \cdot \frac{de^x}{dx}$. It follows that $(b^x)' = (e^{(\ln b)x})' = b^x \cdot \ln b$. Furthermore, since $x^c = \exp(c \ln x)$ for any real c,

$$\frac{dx^c}{dx} = \frac{d \exp(c \ln x)}{d(c \ln x)} \cdot \frac{d(c \ln x)}{dx} = \exp(c \ln x) \cdot \frac{c}{x} = x^c \cdot \frac{c}{x} = cx^{c-1}.$$

This is our familiar power rule extended to all real powers. In fact, the only practical way to compute such x^c converts them to $\exp(c \ln x)$ first.

Rule of 70

If an account earns interest continuously at rate x, with no other deposits or withdrawal, how long T does it take to double? Here the gross return per dollar is $e^{xT} = 2$, so $xT = \ln 2$ or $T \approx 0.693/x$. Often the growth rate is quoted as percentage change per year. In that case, the gross return after t years is $(1+x)^t$ rather than e^{xt} and doubling time is

$$T = \frac{\ln 2}{\ln(1+x)} \approx \frac{1+0.5x}{x} \cdot \ln 2 \approx \frac{0.693 + 0.35x}{x}.$$

For percentage rates in the single digits, the numerator ranges from 0.70 to 0.72. This yields the so-called "rule of 70" or the "rule of 72".

These rules are easy to apply and help to illustrate the amazing nature of exponential growth. Imagine that one of your ancestors saved the equivalent of $10 two thousand years ago, earned 1.0% per year ever since, and passed it down to you. How rich would you be? At first glance, not very. Interest starts at ten cents a year. Each doubling takes nearly 70 years. Still, that allows nearly 29 doublings and yields $4.4 billion.

While inheritances of that magnitude are rare, modern economic growth outstrips previous imagination. China's economy reportedly grew 500% over the past 30 years, which averages to a doubling every decade. In contrast, before 1750 few economies averaged more than 0.2% annual growth, of which most translated into population growth rather than higher living standards.

Similar rules apply to proportionate decay. We replace doubling time with halving times and shift the rule of $69-72$ to the "rule of $66-69$". The shift reflects the x turning negative in the numerator $0.693 + 0.35x$.

If not replenished, radioactivity fades at a constant exponential rate that depends on the type of atom. For example, radioactive carbon-14 decays at rate of 1.22% per century and halves in 5730 years. Cosmic rays striking air continually re-create carbon-14, which plants absorb

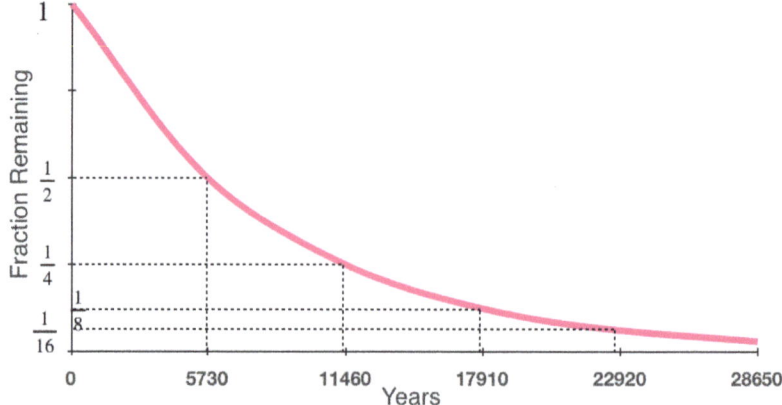

during photosynthesis and animals absorb from plants. After death, internal carbon-14 decays with no new absorption. Analysis has allowed archaeologists to date discoveries as old as 50,000 years.

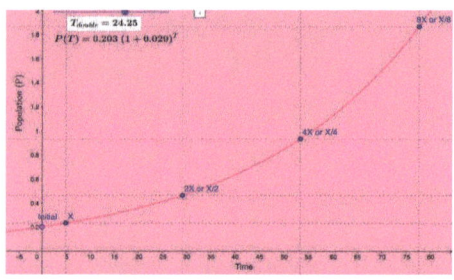

This GeoGebra activity "Doubling Time or Half Life" will let you play with various rates of growth or decay and various benchmarks.

6.3 PROPORTIONAL CHANGE

Rockets

Rockets propel forward by burning fuel and expelling exhaust gases backward. Carried fuel makes the rocket heavier, so more fuel is needed for a given acceleration. How much more?

Let $m(t)$ denote the rocket's total mass at time t, $v(t)$ the rocket's velocity relative to the observer, $-\Delta m = m(t) - m(t + \Delta t)$ the mass of gases expelled in some short interval Δt, and $-w$ the velocity of expelled gases relative to the rocket. The latter is nearly constant since it is determined by the chemical properties of the fuel, the nature of the explosion and the nozzles that project the exhaust.

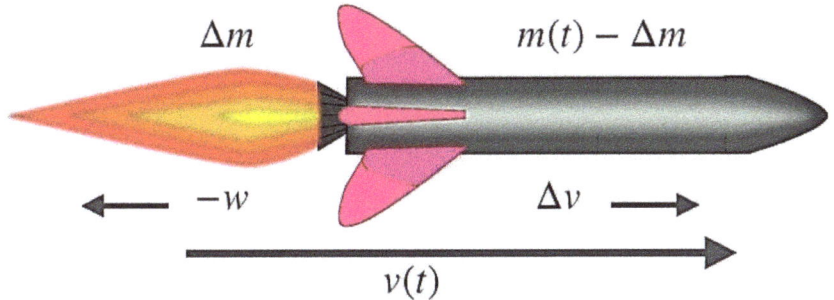

Dividing the changes by Δt and taking the limit, we see that the force propelling the rocket forward is mv' while the exhaust force is wm'. By Newton's Second and Third Laws, these must sum to zero. Applying the chain rule, $\frac{dv}{dt} = -\frac{w}{m} \cdot \frac{dm}{dt} = -w \frac{d \ln m}{dt}$. Integrating both sides from time 0 to T and letting subscripts denote the time,

$$v_T - v_0 = w \ln \frac{m_0}{m_T} \quad \Leftrightarrow \quad \frac{m_0}{m_T} = \exp\left(\frac{v_T - v_0}{w}\right).$$

This is known as the rocket equation, often with the qualifier "ideal" since it ignores external forces.

When the desired velocity boost Δv is several times exhaust velocity w, over 95% of initial mass m_0 must be fuel. Doubling Δv or halving w roughly squares the need for fuel. Consider a rocket using propellants with $w = 4.50$ km/s. To achieve escape velocity $v_T = 11.2$ km/s from a stationary start, $m_0/m_T = \exp(11.2/4.5) \approx 12.0$, which makes fuel nearly 92% of m_0. If every 9 kg of fuel requires at least 1 kg of tanks and engine, the rocket can never escape Earth's gravity.

Multistage rockets improve efficiency. Suppose the stage 1 rocket contains 90% of m_0. If its mass is 90% fuel, that fuel comprises 81% of the whole, for a maximum $\Delta v = w \cdot \ln(100/19) \approx 7.47$ km/s. Suppose the stage 2 rocket contains 90% of the remaining mass, with the same fuel ratio as stage 1. The final payload is $10\% \times 10\% = 1\%$ of m_0. Maximum Δv in stage 2 is the same as in stage 1, for a final velocity of 14.94 km/s, well above escape velocity from Earth.

CHAPTER 6

Cooling

Heat amounts to the random jiggle of molecules. The more they jiggle, the hotter it is. Place a hot object in a colder environment and its molecules will gradually lose their extra energy through conduction.

Let $x(t)$ denote the object's temperature at time t. Newton's Law of Cooling says that the cooling rate $x'(t)$ is directly proportional to the gap $y = x - E$ between $x(t)$ and the temperature $E(t)$ of the environment. Here we will assume E is constant. Then $y' = x' = -by$ for some cooling parameter b. If the environment is massive enough, we can ignore the impact of the hot object and treat E as constant. To solve, we can apply either of the following methods:

- Formally, treat t as an inverse function of y. Integrate $\frac{dt}{dy} = \frac{-1}{by}$ from $y_0 = y(0)$ to $y = y(t)$ to obtain $t = -\frac{1}{b}\ln y + \frac{1}{b}\ln y_0$.
- Informally, treat dy and dt as separate algebraic symbols, rearrange the equation to $\frac{dy}{y} = -b\,dt$, and integrate both sides.

In either case we obtain $\ln y = \ln y_0 - b$, which implies $y = y_0 e^{-b}$. Rewriting in terms of $x = y + E$,

$$x(t) = E + (x_0 - E)e^{-bt}.$$

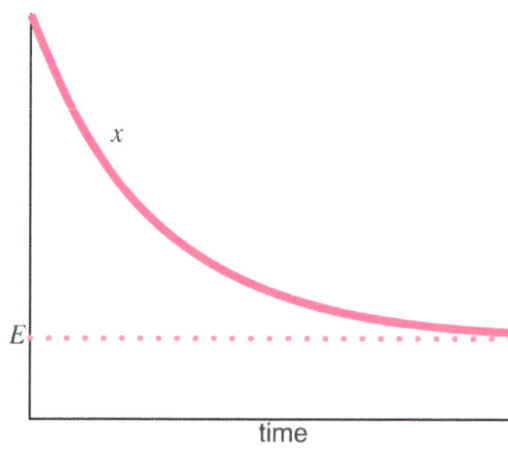

The solution decomposes x into two parts: the long-term stable E and a gap that shrinks exponentially at rate b. This is common behavior in systems approaching equilibrium. In principle, x never fully reaches E. In practice, random fluctuations blur tiny gaps.

> Suppose a bottle of juice left outside on a hot day at $x_0 = 36°C$ is placed in a refrigerator at $E = 4°C$, where it cools in 80 minutes to $20°C$. This halves y for a cooling rate of $b = (60/80)\ln 2 \approx 0.52$ per hour. In another 80 minutes y should halve again, chilling the bottle to $12°C$. For more applications along these lines, see the Chilled Drink Calculator at www.omnicalculator.com/food/chilled-drink.

Air Drag

When an object flies through air, it has to move air molecules out the way. The air's resistance is called drag. At low speeds, streamlined objects part the air smoothly. The corresponding drag force is proportional to the number of molecules moved, which in turn is proportional to object velocity $v(t)$. At high speeds, flying objects cause turbulence. The drag force from the extra jostling is roughly proportional to surface area times $v(t)$ squared. In either case, $v(t)$ will converge to a stable terminal velocity v_∞ given a constant offsetting force like gravity.

If this drag is the only force affecting v, the equation of motion is $v' = -bv$ for some constant b. If the object falls straight down, gravity g opposes drag; this modifies v' to $g - bv(t)$. Setting $v' = 0$ indicates $v_\infty = g/b$. Defining $y = v - v_\infty$, $y' = -by$ just like in the cooling equation, with solution $y = y_0 e^{-b}$ or

$$v(t) = v_\infty + (v_0 - v_\infty)e^{-b}.$$

Defining $w = v/v_\infty$, $w' = b(1-w)$ for linear drag while $w' = b(1-w^2)$ for quadratic drag. The latter multiplies w' by an extra $1+w$ that equals 1 when $v = 0$ and converges to 2. Hence falling objects approach v_∞ nearly twice as fast under quadratic drag as under linear drag.

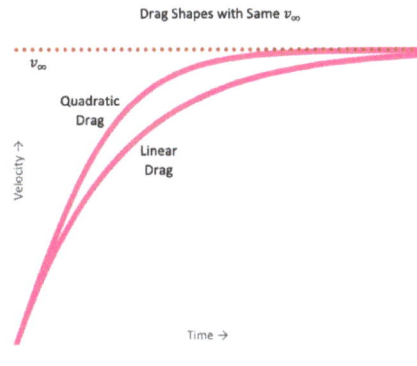
Drag Shapes with Same v_∞

CHAPTER 6

Rewriting the quadratic drag equation in inverse form and integrating,

$$t = \frac{1}{b}\int_{w_0}^{w(t)} \frac{dw}{1-w^2}.$$

To evaluate the integral, note that $\frac{1}{1-w^2} = \frac{½}{1+w} + \frac{½}{1-w}$. Since $1/(1+w)$ and $1/(1-w)$ integrate to $\ln(1+w)$ and $-\ln(1-w)$ respectively, we obtain $t = \frac{1}{2b}\ln\left(\frac{1+w}{1-w}\right)\Big|_{w_0}^{w(t)}$. When $v_0 = 0$, this implies

$$2bt = \ln\left(\frac{1+w}{1-w}\right)$$

$$e^{2bt} = \frac{1+w}{1-w}$$

$$1-w = e^{-2bt} + we^{-2bt}$$

$$1-e^{-2bt} = w\left(1+e^{-2bt}\right)$$

$$v(t) = v_\infty w(t) = v_\infty \frac{1-e^{-2bt}}{1+e^{-2bt}}.$$

Don't memorize these answers. Anyone who needs them can easily look them up. Do trace through the methods. They show how to convert models of nudges into testable hypotheses and useful predictions.

Present Value

When kids first learn about money, they can be easy to fool. They might trade one quarter for ten pennies or one twenty-dollar bill for ten ones. It takes time to appreciate the differences in value and learn to convert money into common units for comparison.

Finance poses a similar challenge to adults. One dollar today is not the same as one dollar next year. We can spend today's dollar now without waiting. We can save or invest today's dollar for extra returns, which come with various risks.

These reasons are closely related. Banks pay interest because people don't like to wait for their money. People don't like to wait in part because of opportunities to save or invest their money for risky profit. In short, future dollars are generally worth less than present dollars.

For a basic analysis, let us ignore risk and assume that all interest is paid continuously at rate r. Suppose we can choose between receiving value V_0 today at time 0 or value V_t at time t. How should we decide?

Let's convert these values to common units. If we save V_0 dollars today we can receive $V_0 e^{rt}$ at t, which we can compare to V_t. Alternatively, we can calculate how much we need to save today to return V_t at t. That amount is $V_t e^{-rt}$, which we can compare to V_0. Either method gives the same ranking. However, since our natural baseline is "now", we usually focus on present values.

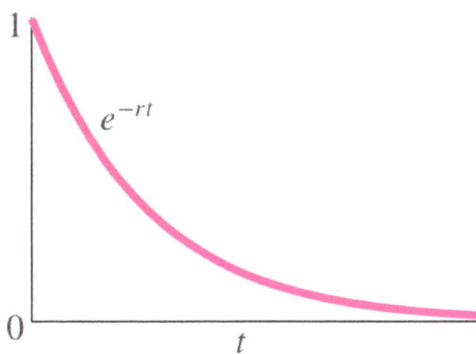

Present value multiplies value at time t by $e^{-rt} < 1$, a procedure known as discounting. The net present value or NPV of future claims sums their discounted values (which for money owed are negative).

If the expected stream of payments is sufficiently smooth, we can approximate NPV using integration. The simplest example is a unit perpetuity: a bond that pays 1 per short period forever. While total payments are unbounded, the NPV is finite:

$$\int_0^\infty 1 \cdot e^{-rt} dt = \left(-e^{-rt}/r\right)\Big|_0^\infty = 0 - (-1/r) = 1/r.$$

In other words, a perpetuity that pays out $r/365$ per day is fairly priced at 1. Not convinced? Compare it to a bank deposit of 1 that we withdraw all the interest from but never the principal. As it pays the same non-risky stream, it deserves the same value.

For a more challenging example, suppose the perpetuity pays out a fraction g/r of interest in new perpetuities rather than cash. That's like holding a bank deposit forever and redepositing a fraction g/r of the interest. While the payments grow at exponential rate g, the NPV still equals 1. Here are two ways to confirm that:

- In an instant dt, a perpetuity valued at P pays $(r-g)dt$ in interest and gdt in new perpetuities worth $Pgdt$. The net return per instant is $r-g+Pg$, which will match Pr if and only if $P=1$.

- If the stream of payments grows exponentially at rate g from an initial base of $r-g$, NPV is $(r-g)\int_0^\infty e^{gt}e^{-rt}dt = (r-g)\int_0^\infty e^{(g-r)t}dt = 1$.

Debt Bubbles

Now for something trickier. Suppose g exceeds r for the perpetuity. The NPV is infinite, so lenders lend regardless of price. Borrowers love it too. By borrowing new money at growth rate g and repaying old debts at rate r, they harvest current income at rate $g-r$. However, debt keeps mounting, potentially beyond what the debtor can directly repay. This is known as a debt bubble and usually ends badly.

Currently, the developed world is trying to capture the win-win aspects of debt bubbles without triggering collapse. One marker is the U.S. debt to GDP ratio, which has risen to levels unprecedented in peacetime. There is no sign yet of reversal.

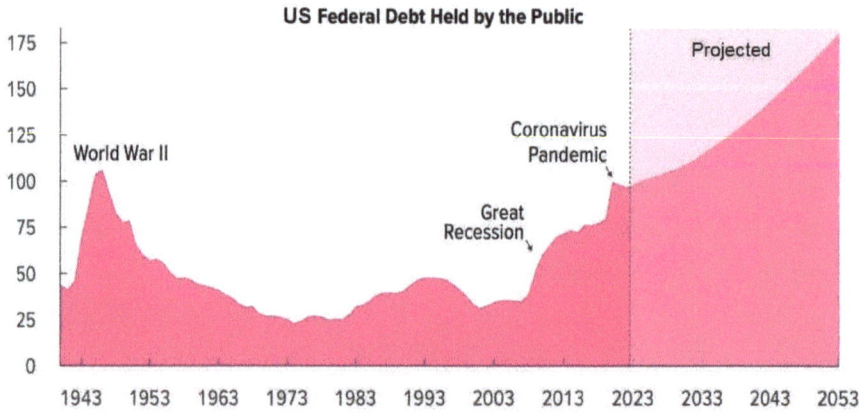

6.4. FLUCTUATING GROWTH

Damped Oscillations

The previous section modeled growth that either stays a fixed proportion of the value we're tracking or fades monotonically over time. This section uses exponentials to model fluctuating growth. Let's start with a spring subject to friction. Absent friction, as we recall from Chapter 5.4, the spring's extension x satisfies $x'' = -cx$ for some constant c. Modeling friction as linear in velocity x' modifies that to $x'' = -bx' - cx$.

Experience shows that friction gradually slows oscillation. Might we view x as the product of stable oscillation and exponential decay? Let us try. If $x(t) = y(t)e^{-rt}$, then

$$x' = y'e^{-rt} - rye^{-rt} \text{ and } x'' = y''e^{-rt} - 2ry'e^{-rt} + r^2ye^{-rt}.$$

Substitute into $x'' = -bx' - cx$, factor out e^{-rt}, and rearrange to yield

$$y'' = (2r - b)y' + (-r^2 + br - c)y.$$

If we set $r = \frac{1}{2}b$, the term in y' drops out, leaving $y'' = sy$ where $s = \frac{1}{4}b^2 - c$, $y(0) = x_0 e^{r0} = x_0$ and $y'(0) = x'(0)e^{r0} + rx(0)e^{r0} = v_0 + \frac{1}{2}bx_0$. This has three types of solutions depending on the sign of s:

Underdamping: For $s < 0$, $y'' = -|s|y$ is the frictionless spring equation studied before. The solution takes the form

$$y(t) = A\cos qt + B\sin qt \text{ for } q = \sqrt{|s|} = \sqrt{|\tfrac{1}{4}b^2 - c|}.$$

The initial conditions require $A = x_0$ and $qB = v_0 + \frac{1}{2}bx_0$. Hence

$$x(t) = \left(x_0 \cos qt + w_0 = \frac{v_0 + \frac{1}{2}bx_0}{q} \sin qt\right) e^{-\frac{1}{2}bt}.$$

With more work we can rewrite this as $x(t) = Ce^{-\frac{1}{2}bt}\cos q(t - \tau)$ for properly chosen C and τ. While the formula isn't simple, it is neat how smoothly we can deduce it and how much insight it offers. The factor $\frac{1}{2}b$ indicates how fast the maximum amplitude C decays. The factor q indicates the pace of oscillation and how friction slows it.

CHAPTER 6

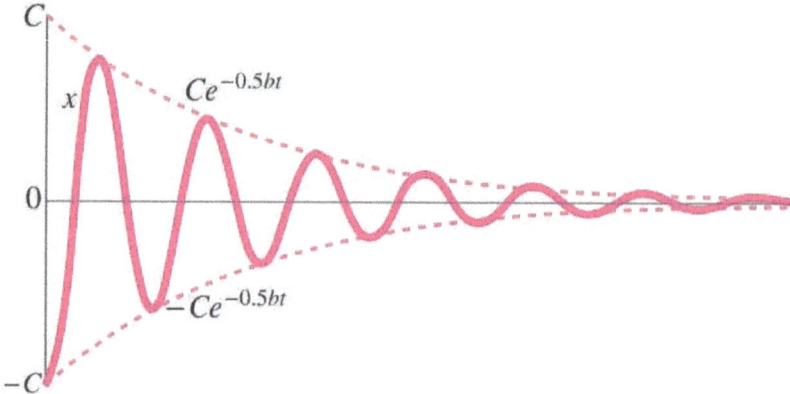

Overdamping: For $s > 0$, the general solution is $y(t) = Ee^{qt} + Fe^{-qt}$ or

$$x(t) = Ee^{(q-\frac{1}{2}b)t} + Fe^{-(q+\frac{1}{2}b)t}.$$

Since $q = \sqrt{b^2/4 - c}$, both exponents in x are negative. The solution mixes two types of ordinary exponential decay. The initial conditions $E + F = x_0$ and $q(E - F) = v_0 + \frac{1}{2}bx_0$ imply

$$E = \left(\frac{1}{2} + \frac{b}{4q}\right)x_0 + \frac{1}{2q}v_0 \quad \text{and} \quad F = \left(\frac{1}{2} - \frac{b}{4q}\right)x_0 - \frac{1}{2q}v_0.$$

If started from rest, the spring reverts monotonically toward the origin. If pushed hard enough, it overshoots the origin once before it reverts.

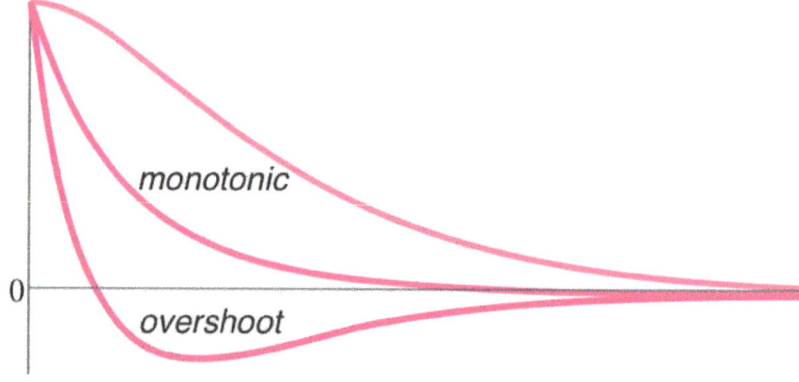

Exact Damping: For $s = 0$, $y'' = 0$, so $y(t) = x_0 + (v_0 + \frac{1}{2}bx_0)t$ and

$$x(t) = x_0 e^{-\frac{1}{2}b} + (v_0 + \frac{1}{2}b_0)te^{-\frac{1}{2}b}.$$

The multiplier of t on the second term has no counterpart in overdamping. However, the equation implies similar behavior. Started from rest, an exactly damped spring reverts monotonically to the origin. Given a push, it overshoots at most once. How can we be sure? Solve $x'(t) = 0$ for stationary points. Since x' takes the form $(v_0 + Ct)e^{-\frac{1}{2}bt}$, this has at most one positive solution and none for $v_0 = 0$.

The GeoGebra activity "Damped Oscillations" lets you explore these varied behaviors as well as the case $c < 0$ which implies long-term exponential growth.

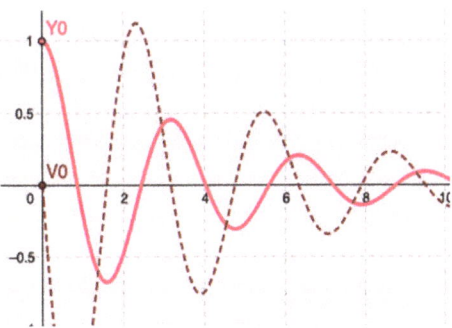

Logistic Growth

Suppose that y people in a large population P contracts a contagious disease. On average, each infected person spreads the disease to b others per day, of which a fraction $x = y/P$ are already infected. The daily contagion rate is $b(1 - y/P)$ per spreader or $y' = by(1 - y/P)$ overall. Hence

$$x' = bx(1-x).$$

This is known as logistic growth. The graph of x' versus x is quadratic pointing down. It vanishes at 0 and 1 with a peak at $x = \frac{1}{2}$. This divides logistic growth into three phases:

- nearly exponential growth with x' approaching bx.
- nearly linear with $x' \approx \frac{1}{4}b$.
- nearly exponential slowing with x' approaching $b(1-x)$.

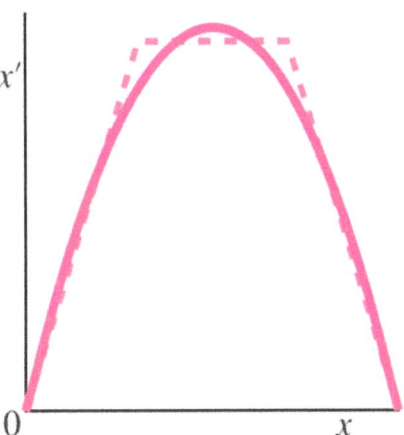

CHAPTER 6

To solve, rewrite as $b\,dt = \dfrac{dx}{x(1-x)} = \dfrac{dx}{x} + \dfrac{dx}{1-x}$ and integrate to obtain $b + \ln C = \ln x - \ln(1-x) = \ln \dfrac{x}{1-x}$. Taking antilogarithms of both sides, $\dfrac{x}{1-x} = Ce^{bt}$. Rearranging,

$$x(t) = \dfrac{Ce^b}{Ce^b + 1} = \dfrac{1}{1 + De^{-bt}} \quad \text{for} \quad D = \dfrac{1}{C} = \dfrac{1-x_0}{x_0}.$$

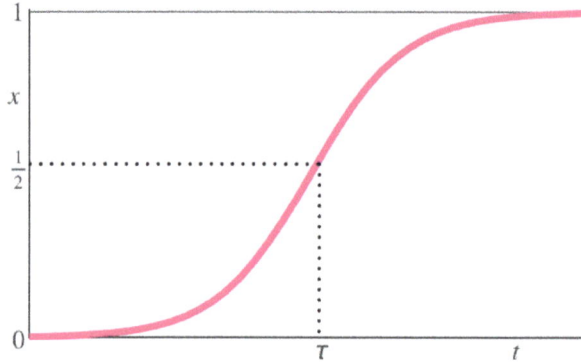

This is an S-shaped function known as the logistic. It ranges from 0 to 1. Here $\tau = (\ln D)/b$ marks the inflection point.

The logistic represents proportional growth subject to a topping-out constraint. It arises in a variety of contexts. For example, fish y breeding at rate by on a fish farm might be constrained by the pond's carrying capacity Y and by harvesting at a fraction c of the breeding rate. If net survival is linear in crowding, growth will take the form

$$y' = b\left(1 - \dfrac{y}{Y}\right) - cb = b(1-c)y\left(1 - \dfrac{y}{1-cY}\right).$$

Defining $x = \dfrac{y}{(1-c)Y}$ converts this into logistic growth at rate $b(1-c)$.

People

Exponentials are so appealing for modeling growth that sometimes we project them where we shouldn't. As we've just seen, logistic growth is nearly impossible to distinguish in its early stages from purely exponential growth. Yet the two forms imply sharply different long-term behavior. Fitting a good-looking curve doesn't mean we've identified the driving forces.

Human population trends illustrate how important this can be. World population reached one billion in about 1800, having taken 300 years to double. Over the next 125 years it doubled again. The next doubling after that took less than 50 years. Is that not exponential growth at an ever-quickening pace?

Yet in recent decades world population growth has sharply decelerated. Annual growth from 1950 through 1975 averaged nearly 2.0%. Now it is less than 1.0% with further slowdown expected. The shift has three main causes:

- People are living longer but old people rarely have kids, so the birth rate per adult falls.
- Women are having fewer children, so birth rates per potential mother are falling.
- Mothers are having children later in life, which slows down population increase.

However, poor countries are significantly lagging in the shift.

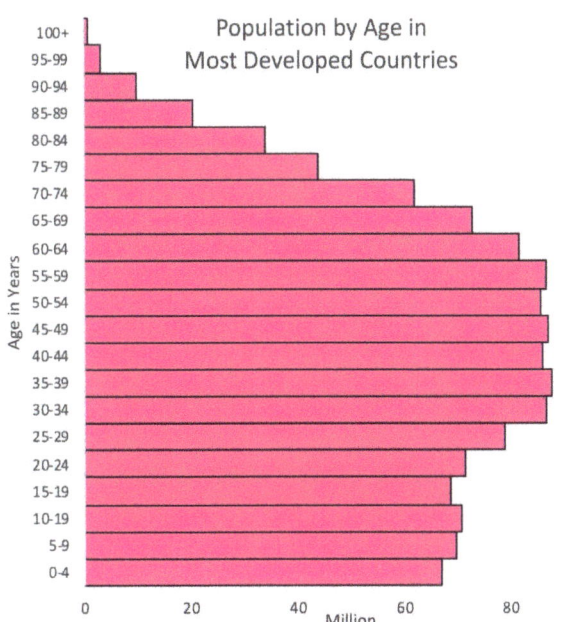

Population pyramids, which subdivide by age group, help make this clear. Here is the pyramid for the most developed countries. The number of children has shrunk by 25%, which bodes fewer active workers to support the old.

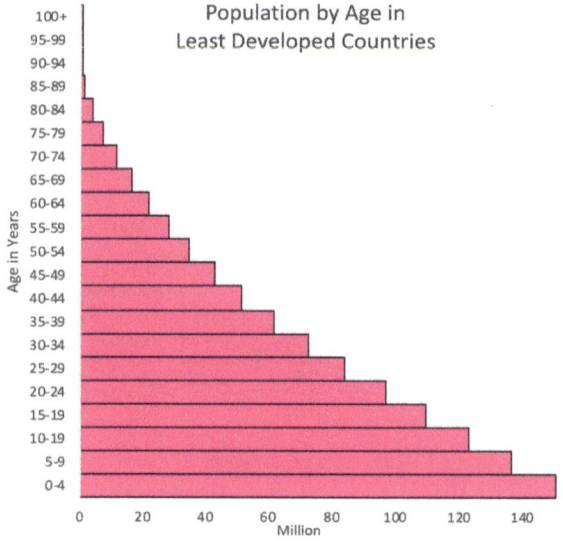

Here is the pyramid for the poorest countries. Their future workers—over twice as many as in the richest countries—are struggling to obtain the opportunities they need to flourish.

Predators and Prey

Both damped oscillations and logistic growth converge to stable states. However, circular orbits show that not all equilibria are static. In fact, some equilibria look quite irregular. The most striking example is a stylized predator-prey model, known as Lotka-Volterra after its inventors.

In this model, prey x have plenty of food to munch on but get munched by predators y. Apart from predation, prey increase and predators decrease in direct proportion to their numbers. Predation depletes prey and boosts predators (via food for young) in direct proportion to their potential encounters xy. The dynamics can be summarized as $x' = Ax - Bxy$ and $y' = Cxy - Dy$ for constants A, B, C, D.

The resulting behavior is strange in many ways:

- The mix of x and y returns infinitely often to the starting mix.
- The behavior is cyclical, with constant time per cycle.
- x and y tend to fluctuate wildly through the cycle.
- In most cycles, x or y occasionally get so scarce that a few chance deaths would cause extinction.

The GeoGebra activity "Lotka-Volterra Model" lets you play with the parameters.

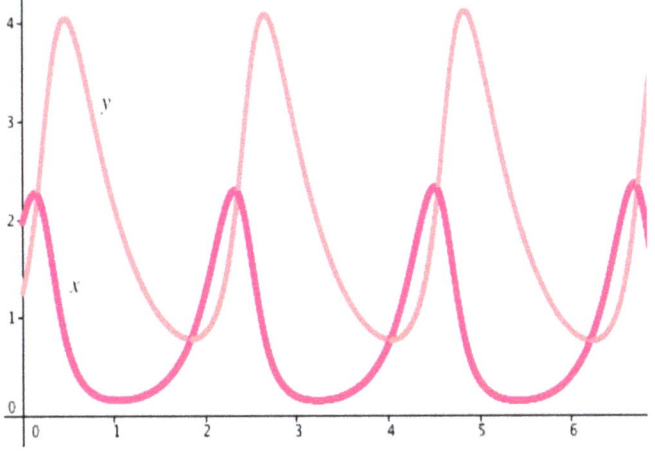

For more insight, divide the y' equation by the x' equation and apply the chain rule to obtain $\dfrac{y'}{x'} = \dfrac{dy}{dx} = \dfrac{y(Cx-D)}{(A-By)x}$, which we can rearrange to

$$\dfrac{A-By}{y}dy = \dfrac{Cx-D}{x}dx.$$

Then integrate both sides to see that $A\ln y - By = Cx - D\ln x + \ln K$ for some constant K, which implies

$$y^A e^{-By} e^{-Cx} x^D = K.$$

Graphed in an x, y-plane, this describes an oblong orbit for every K. The (x, y) mix moves counterclockwise in its orbit.

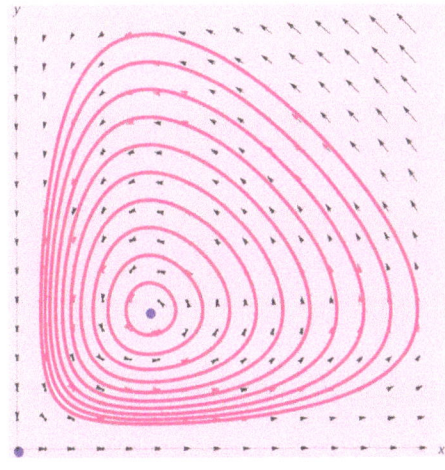

The model's neatness misleads. Competition between prey (so that A decreases with x) or predators (so that C decreases with y) will make the cycles spiral toward a single equilibrium. Still, since the equilibrium will change over time, the spirals may seem never-ending.

7

Circle Functions

Complex numbers combine real numbers with the square roots of negative numbers. While they take some getting used to, they simplify the study of periodic motion. Calculus with complex numbers is remarkably easy and surprisingly powerful.

7.1. COMPLEX NUMBERS

Square Root of -1

As we have seen, a quadratic equation $y = ax^2 + bx + c$ describes a parabola. The parabola crosses the x-axis twice if $b^2 > 4ac$, touches it once if $b^2 = 4a$, and never reaches it if $b^2 < 4ac$. "Never reaches" means that $ax^2 + bx + c = 0$ has no real number solution.

When equations can't be solved with existing numbers, mathematicians sometimes invent new types of numbers to solve them. Rational numbers solve equations like $3x = 5$ that whole numbers cannot solve. Real numbers solve equations like $x^2 = 2$ that rational numbers cannot solve. Imaginary numbers solve equations like $x^2 = -1$.

Imaginary numbers pose two special challenges. The first is their name. We never see, hear, taste, smell or touch a pure number. Every whole number is empty, every rational number is torn between two whole numbers, and no real number can really be distinguished from its closest neighbors. The only good reason for singling out some numbers as "imaginary" is that they cannot be ordered on a real number scale.

The second challenge is doing math with imaginary numbers. The easiest way defines $i = \sqrt{-1}$ and expresses every other imaginary number as a real multiple b of i. Sums of real and imaginary numbers can be written in the form $z = a + bi$, known as complex numbers.

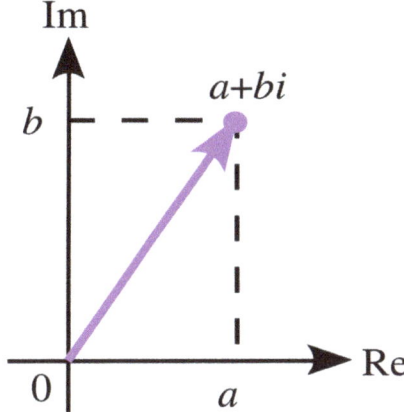

To visualize z, assign it to a point on the plane, with real component $\text{Re}(z) = a$ on the horizontal axis and imaginary component $\text{Im}(z) = b$ on the vertical axis. For more emphasis, link $a + bi$ to a radial vector extended from the origin. This is called an Argand diagram.

The addition rule for complex numbers is

$$(a + bi) + (c + Di) = (a + c) + (b + D)i,$$

where all the coefficients are real and D replaces d so as not to confuse di with a differential. Complex addition works just like its vector counterpart. Without changing the length or direction of the $c + Di$ vector, move its base to $a + b$ and mark the sum at the new head. To subtract, add with negative sign, which reverses the second vector's direction:

$$(a + bi) - (c + Di) = (a - c) + (b - D)i.$$

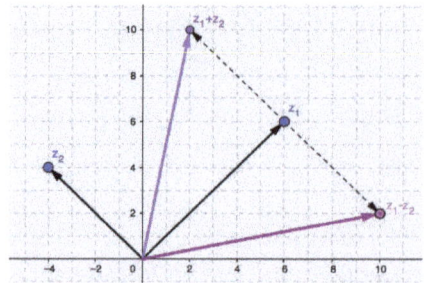

The GeoGebra activity "Complex Addition and Subtraction" demonstrates their operation.

To multiply complex numbers, we apply the distributive rule twice. Since $(a+b\,)c = a\,c + b\,c\,i$ and $(a+b\,)Di = aD\,i + bD\,i^2 = aD\,i - bD$,

$$(a+bi)(c+Di) = (ac - bD) + (aD + bc)i.$$

For an interesting special case, multiply $a + bi$ by what is known as its "conjugate" value $a - bi$:

$$(a+b\,)(a-b\,) = a\,a - b(-b) + (-ab + ab\,)i = a^2 + b^2.$$

Hence $\dfrac{c+Di}{a+b} = \dfrac{(a-bi)(c+Di)}{a^2+b^2}$. Presto! We have just turned complex division into complex multiplication and real division.

Complex multiplication shows several regularities, including:

- **Vector lengths multiply:** The vectors $a+bi$ and $c+Di$ have squared lengths a^2+b^2 and c^2+D^2. The squared length of their product simplifies to $(a^2+b^2)(c^2+D^2)$.

$$(ac-bD)^2 + (aD+bc)^2$$
$$= a^2c^2 - 2abcD + b^2D^2$$
$$+ a^2D^2 + 2abcD + b^2c^2$$
$$= (a^2+b^2)(c^2+D^2)$$

- **Real multiplication retains slant:** Since $(a+bi)c = ac + bci$, real and imaginary components change in equal proportion. A positive c preserves the direction while a negative c reverses it.

- **Imaginary multiplication rotates a quarter turn:** Since $(a+bi)i$ equals $-b + ai$, the real component shifts to the positive imaginary axis while the imaginary component shifts to the negative real axis. The vector rotates 90° counterclockwise. In particular, $i^2 = -1$, $i^3 = -i$, and $i^4 = 1$.

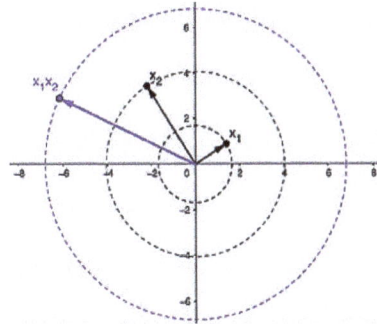

If you experiment with the GeoGebra activity "Complex Multiplication", you will see that it appears to add the angles of vectors it multiplies. We will confirm this shortly.

Complex Exponentials

Let's extend our Maclaurin series definition of exponentials to cover complex numbers too. To confirm that $e^w e^z = e^{w+z}$ for any w and z, note first that

$$e^w e^z = \left(\sum_{m=0}^{\infty} \frac{w^m}{m!}\right)\left(\sum_{n=0}^{\infty} \frac{z^n}{n!}\right) = \sum_{m=0}^{\infty}\sum_{n=0}^{\infty} \frac{w^m z^n}{m! n!}.$$

By the Binomial Theorem, $e^{w+z} = \sum_{k=0}^{\infty} \frac{(w+z)^k}{k!}$ also sums up multiples of $w^m z^n$, although it orders them by $k = m+n$. The coefficient on $w^m z^n$ is $\frac{(m+n)!}{m!n!} \cdot \frac{1}{(m+n)!} = \frac{1}{m!n!}$, which is the same as in the formula above.

By the multiplication rule, adding a real number c to a complex z just rescales e^z by a factor e^c. But what does a purely imaginary exponent do? Let us denote it $i\theta$, where θ ("theta") is real. Here is the Maclaurin series for $e^{i\theta}$:

$$e^{i\theta} = \sum_{k=0}^{\infty} \frac{i^k \theta^k}{k!} = 1 + i\theta - \frac{\theta^2}{2} - i\frac{\theta^3}{3!} + \frac{\theta^4}{4!} + i\frac{\theta^5}{5!} - \cdots.$$

The real component matches $\cos\theta$ from Chapter 5.4 while the imaginary component matches $\sin\theta$. Rearrangement yields what is known as Euler's formula:

$$e^{i\theta} = \cos\theta + i\sin\theta.$$

CHAPTER 7

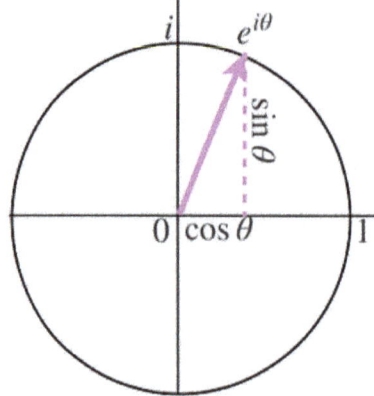

The Maclaurin series for $e^{-i\theta}$ matches $e^{i\theta}$ except that terms in odd powers of θ flip signs. Thus $e^{-i\theta} = \cos\theta - i\sin\theta$. Since $e^{i\theta}e^{-i\theta} = e^0 = 1$,

$$(\cos\theta)^2 - (i\sin\theta)^2 = \cos^2\theta + \sin^2\theta = 1.$$

It follows that $e^{i\theta}$ lies on the unit circle. But where?

For a nudge $\Delta\theta$, the point $e^{i\Delta\theta} \approx 1 + i\Delta\theta$ is about $\Delta\theta$ distant from $e^{i0} = 1$. In the limit, the ratio of arc length to nudge equals 1.

By symmetry, this must be true everywhere on the circle. Hence the arc from e^{i0} counterclockwise to $e^{i\theta}$ has length θ.

Complex Implications

This is a wondrous result. Even though $e^{i\theta}$ is defined by infinite complex Taylor series, we can place it on a circle as clearly as we can place a real number on a line. That is why $e^{i\theta}$ and its $\cos\theta$ and $\sin\theta$ components are called circle functions. Interesting properties include:

- **Euler's identity:** Since an arc of length π covers a half-circle, $e^{\pi i} + 1 = 0$. This famous equation links the five most important constants in higher math: $0, 1, i, \pi$ and e.

- **Radian measure:** Every arc length on the unit circle can be viewed as measuring the associated angle. Its units are called radians. To convert to degrees, multiply by $180/\pi$.

- **Cyclic repetition:** As every arc of length 2π returns to the starting point, $e^{2\pi k i} = 1$ for every integer k. It follows that $\cos\theta = \cos(\theta + 2\pi k)$ and $\sin\theta = \sin(\theta + 2\pi k)$.

- **Full mapping:** Every complex $z = a + bi$ can be expressed as
  $$re^{i\theta} = r(\cos\theta + i\sin\theta),$$
  where $r = \sqrt{a^2 + b^2}$ denotes the distance from the origin and θ satisfies $\dfrac{\sin\theta}{\cos\theta} = \dfrac{b}{a}$.

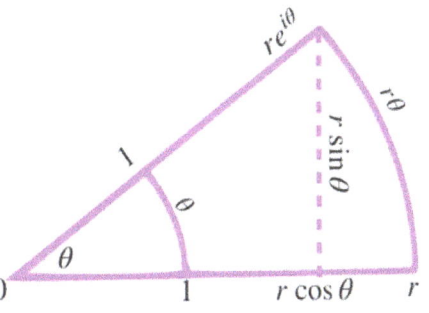

- **Phase shift (rotation):** The cosine function matches the sine function a quarter turn later: $\cos\theta = \sin(\theta + \tfrac{1}{2}\pi)$.

- **Multiplication as stretching and rotation:** The product of two complex numbers $re^{i\theta}$ and $se^{i\tau}$ is $rse^{i(\theta+\tau)}$. Radial distances multiply and angles add.

- **Cosines and sines of sums:** Matching up the imaginary and real components of $e^{i(\theta+\tau)} = e^{i\theta}e^{i\tau}$ shows that
  $$\cos(\theta + \tau) = \cos\theta\cos\tau - \sin\theta\sin\tau;$$
  $$\sin(\theta + \tau) = \sin\theta\cos\tau + \cos\theta\sin\tau.$$

- **De Moivre's formula:** Since $e^{i(n\theta)} = \left(e^{i\theta}\right)^n$, it follows that
  $$\cos n\theta + i\sin n\theta = (\cos\theta + i\sin\theta)^n.$$

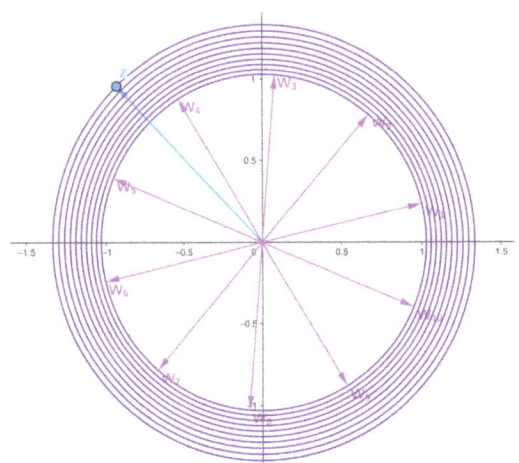

When $\theta = 2\pi k/n$, the left-hand side of de Moivre's formula is 1. The right-hand side implies that the n complex roots of 1 form n equal arcs. The GeoGebra activity "Complex Roots" displays them.

CHAPTER 7

Polar Coordinates

The (x, y) values of a point are known as rectangular coordinates since they measure distances along a rectangle from the origin. Vertical lines of constant x and horizontal lines of constant y intersect at right angles.

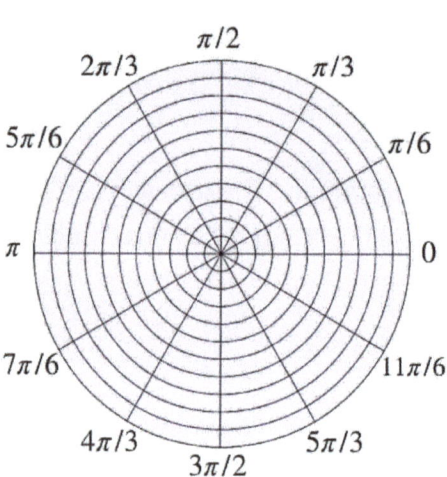

The corresponding (r, θ) values are known as polar coordinates since they measure distance from a pole and rotation from an initial direction. The circles link points with the same r while the rays link points with the same θ. The rays and circles intersect at right angles. Like vertical lines or horizontal lines, the various circles stay fixed distances apart, However, the distances between rays widen in direct proportion to the distance r from the origin.

Rectangular coordinates are most convenient when we shift the origin ("translate") to (t_x, t_y), since we just subtract t_x from x and t_y from y to compensate. Polar coordinates are most convenient when we rotate the baseline direction. For axis rotation τ radians counterclockwise, every θ maps to $\theta - \tau$ with no change in r. If we convert (r, θ) to (x, y), x maps to $x \cos \tau + y \sin \tau$ while y maps to $y \cos \tau - x \sin \tau$. Object rotation by τ is like axis rotation by $-\tau$. If the new axis has slope u relative to the old, then $\cos \tau = 1/\sqrt{1+u^2}$ and $\sin \tau = u/\sqrt{1+u^2}$, and any equation in x and y for the object makes the substitution

$$(x, y) \to \left(\frac{x - uy}{\sqrt{1+u^2}}, \frac{y + ux}{\sqrt{1+u^2}} \right).$$

7.2 COMPLEX CALCULUS

Circular Motion

Calculus treats complex constants just like real constants. In particular, $\frac{de^{i\theta}}{d\theta} = \frac{de^{i\theta}}{d(i\theta)} \cdot \frac{d(i\theta)}{d\theta} = ie^{i\theta}$. For a simple application, let us reconsider circular motion with constant speed v at radius r. Given position $z(t) = re^{i\theta(t)}$, the velocity $v = z' = re^{i\theta} \cdot i\theta'$. Its direction $ie^{i\theta}$ must be perpendicular to the radial vector $e^{i\theta}$, while angular rotation θ' must equal v/r. Acceleration $z'' = v \cdot ie^{i\theta} \cdot i\theta' = -e^{i\theta}v^2/r$ always points to the center with magnitude v^2/r. If the acceleration comes from gravity, which is reciprocal to r^2, then $v^2 r$ will be constant.

Chapter 4.3 derived these results earlier using separate x and y. Differentiating $e^{i\theta}$ unites them. Moreover, the radial vector highlights the logical connection. Each derivative rotates the vector by a quarter turn, which makes acceleration rotate half a turn and point inward.

Like with real numbers, complex integration is just the reverse of differentiation. However, the area under a curve doesn't match its definite integral. When $e^{i\theta}$ moves from $\theta = 0$ to $\theta = \pi$, it traces a unit semi-circle of area $\pi/2$, yet $\int_0^\pi e^{i\theta} d\theta = -ie^{i\theta}\Big|_0^\pi = 2i$.

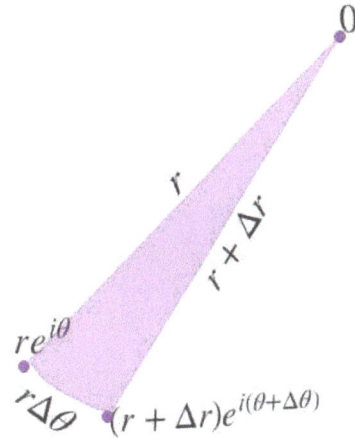

To fix this, we need to think more about nudges in area. When θ changes by $\Delta\theta$, $re^{i\theta}$ swings by an arc of length $r\Delta\theta$. The extra region is nearly triangular with height of about r, base of about $r\Delta\theta$, and area of about $\frac{1}{2}r^2\Delta\theta$.

Hence the integral for area is $\int \frac{1}{2} r^2 d\theta$. For example, when r is constant, $\int_0^{2\pi} \frac{1}{2} r^2 d\theta = \frac{1}{2} r^2 \theta \big|_0^{2\pi} = \pi r^2$ tells us the area of a circle. In contrast, $\int_0^{2\pi} re^{i\theta} d\theta = -ire^{i\theta} \big|_0^{2\pi} = 0$ tells us that the circle ends where it begins.

Kepler's Second Law

Copernicus published his theory of a nearly sun-centered universe in 1543. At first, his theory made little headway. It couldn't explain why we don't fall off Earth or feel its motion. Its presumption of circular orbits with constant speed wasn't fully consistent with the evidence. Its only justification was simpler approximation to what we observe.

Expert consensus didn't turn until, over 75 years later, Kepler offered three ingenious amendments, framed as natural laws:

I. Planets orbit in ellipses, with Sun at one focus.

II. For a given orbit, a chord from Sun to planet sweeps through equal areas in equal time.

III. Across planets, orbital period squared is proportional to the longest radius cubed.

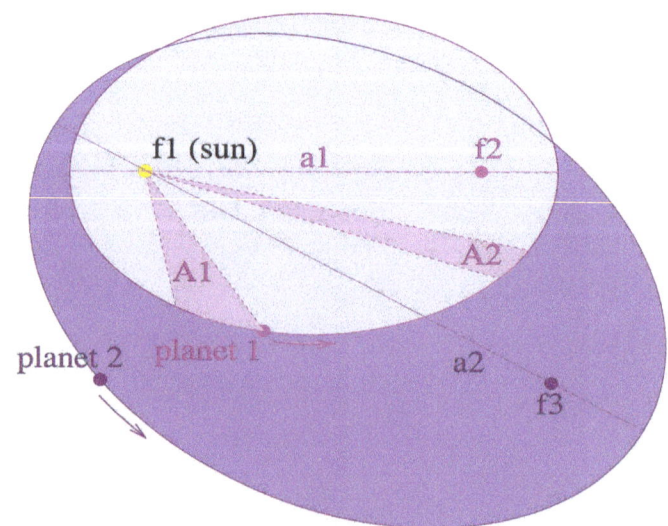

Like Copernicus, Kepler offered no rationale for these laws. He just noted that they worked and exalted the cosmic harmony. Decades more passed until Newton, using calculus, revealed the connection to gravity. The revelation was so elegant and persuasive that old objections withered away.

To present this neatly, let us recalculate motion in complex terms for position $z = re^{i\theta}$. The derivatives with respect to t identify radial components along $e^{i\theta}$ and angular components $ie^{i\theta}$ perpendicular to it:

$$z' = r'e^{i\theta} + re^{i\theta}i\theta'$$
$$z'' = r''e^{i\theta} + i2r'\theta' e^{i\theta} + ir\theta'' e^{i\theta} - r\theta'\theta' e^{i\theta}$$
$$= \left(r'' - r\theta'\theta'\right)e^{i\theta} + \left(2r'\theta' + r\theta''\right)ie^{i\theta}$$

Assume gravity comes from the origin. As it provides radial acceleration only, the transverse component must vanish:

$$2r'\theta' + r\theta'' = 0.$$

To solve, multiply by r to obtain $2rr'\theta' + r^2\theta'' = 0$. Since the left-hand side is the derivative of $r^2\theta'$, $r^2\theta'$ must equal some constant K. Now think about what that means. Let $A(t) = \int_{\theta(0)}^{\theta(t)} \tfrac{1}{2} r^2 d\theta$ denote the area swept out by the radial vector from time 0 to t. Applying the Fundamental Theorem and the chain rule,

$$A'(t) = \tfrac{1}{2} r^2 \theta' = \tfrac{1}{2} K.$$

Hence, the rate A' at which area is swept out is constant. We have proved Kepler's Second Law.

Elliptical Orbits

From Chapter 4.3, when Sun of mass M and planet of mass m tug on each other, net acceleration from gravity is $-G(M+m)/r^2$. Equate this to radial acceleration $r'' - r\theta'\theta'$ and substitute $\theta' = K/r^2$ to obtain

$$r'' - \frac{K^2}{r^3} = \frac{-G(M+m)}{r^2}. \qquad (EQ^*)$$

This looks both beautiful and ugly. Beautiful, because it neatly ties together r'' and powers of r. Ugly, because the powers rule out familiar solutions.

Since orbits look elliptical, let us derive an elliptical formula for r and check whether it satisfies EQ^*. For an ellipse centered at the origin with largest radius a, we learned in Chapter 3.2 that the distance from focus $(0,c)$ to point (x,y) is $a - (c/a)x$. That is the r we need. Then make two substitutions:

- $x = c + r\cos\theta$, which sums the distance from origin to focus and the horizontal component of r.
- $\varepsilon = c/a$, a number between 0 and 1 that describes how squashed or "eccentric" the ellipse is.

Hence $r = a - \varepsilon x = a - \varepsilon(\varepsilon a + r\cos\theta)$ with initial condition $r(0) = r_0$. The solution is

$$r = \frac{a(1-\varepsilon^2)}{1+\varepsilon\cos\theta} = \frac{r_0(1+\varepsilon)}{1+\varepsilon\cos\theta}.$$

This implies radial velocity and acceleration in orbit of

$$r' = \frac{-r_0(1+\varepsilon)}{(1+\varepsilon\cos\theta)^2} \cdot \frac{d\varepsilon\cos\theta}{d\theta} \cdot \frac{d\theta}{dt} = \frac{\varepsilon\sin\theta}{r_0(1+\varepsilon)} \cdot r^2\theta' = \frac{\varepsilon K \sin\theta}{r_0(1+\varepsilon)}$$

$$r'' = \frac{\varepsilon K \cos\theta}{r_0(1+\varepsilon)} \cdot \theta' = \frac{\varepsilon K^2 \cos\theta}{r_0(1+\varepsilon)r^2}$$

If we plug this r'' into EQ^* and multiply both sides by $-r^2$, we obtain

$$-\frac{\varepsilon K^2 \cos\theta}{r_0(1+\varepsilon)} + \frac{K^2(1+\varepsilon\cos\theta)}{r_0(1+\varepsilon)} = \frac{K^2}{r_0(1+\varepsilon)} = G(M+m),$$

which many combinations of K, r_0, and ε can satisfy.

Kepler's First Law

To see what other solutions EQ^* can have, it helps to define a new variable $s(\theta) = s(\theta(t)) = 1/r(t)$. Differentiating,

$$r' = \frac{ds^{-1}}{dt} = \frac{-1}{s^2} \cdot \frac{ds}{d\theta} \cdot \frac{d\theta}{dt} = -r^2 s'\theta' = -Ks'.$$

Hence $r'' = -Ks''\theta' = -K^2 s^2 s''$. Substitution converts EQ^* into

$$-K^2 s^2 s'' - K^2 s^3 = -G(M+m)s^2.$$

Divide by $-K^2 s^2$ to obtain

$$s'' + s = S \quad \text{for} \quad S = \frac{G(M+m)}{K^2}.$$

This form is much simpler than EQ^*. It has a constant solution $s = S$, which implies circular motion with $r = 1/S$. Any other $s = q + S$ is a solution if and only if $q'' + q = (s-S)'' + s - S = s'' + s - S = 0$.

As we saw in Chapter 5.4, $q'' = -q$ is an equation for an undamped spring. The solution is $q = B\cos(\theta + C)$ for some real B and C, with maximum value $S + |B|$ for s. To simplify, let us set $\theta = t = 0$ where r is minimized. This requires initial motion perpendicular to Sun, $K = r_0 v_0$, $B > 0$, and $C = 0$. Then

$$r = \frac{1}{s} = \frac{1}{S + B\cos\theta} = \frac{S^{-1}}{1 + BS^{-1}\cos\theta} = \frac{r_0(1+\varepsilon)}{1 + \varepsilon\cos\theta},$$

where $\varepsilon = BS^{-1} > 0$. If $\varepsilon \geq 1$, the implied r is infinite or negative at $\theta = \pi$. Hence all closed (repeating) orbits have $0 < \varepsilon < 1$, which as we have seen describes ellipses.

CHAPTER 7

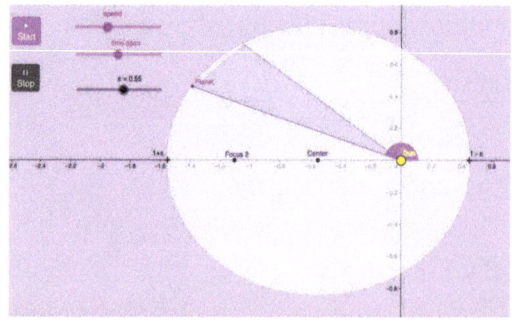

The GeoGebra activity "Kepler's First Two Laws" simulates their operation.

Kepler's Third Law

By Kepler's Second Law, the radial vector sweeps out area $\tfrac{1}{2}KT$ in orbital period T. By Kepler's Second Law, this must match the ellipse area πab calculated in Chapter 3.3, so $T = 2\pi ab/K$. Note that

$$b^2 = a^2 - c^2 = a^2(1-\varepsilon^2),$$

since the distances to $(0,b)$ from foci $(c,0)$ and $(-c,0)$ must sum to $2a$. Also, $K^2 = G(M+m)S^{-1}$ and $S^{-1} = r_0(1+\varepsilon) = a(1-\varepsilon^2)$. Hence

$$\frac{T^2}{a^3} = \frac{4\pi^2 a^2 b^2}{a^3 K^2} = \frac{4\pi^2 a(1-\varepsilon^2)}{G(M+m)S^{-1}} = \frac{4\pi^2}{G(M+m)}.$$

This is Kepler's Third Law, with one caveat. Since $M+m$ varies by planet, T^2 won't be exactly proportional to a^3. However, since even huge Jupiter has only one-thousandth the mass of the Sun and most planets are far smaller, differences are hard to detect. Usually $M+m$ is simplified to M.

Gravity Slingshots

Let's look more closely at solutions to EQ^* with $\varepsilon \geq 1$. They arise whenever $r_0 v_0^2 = K^2/r_0 \geq 2G(M+m)$. In that case the object moves too fast transverse to Sun for solar gravity to permanently capture it. Since $\cos\theta$ is bounded by $-1/\varepsilon$ below and $1/\varepsilon$ above, the flight path never completes a loop. While the path curves around Sun, it straightens out as it recedes.

CIRCLE FUNCTIONS

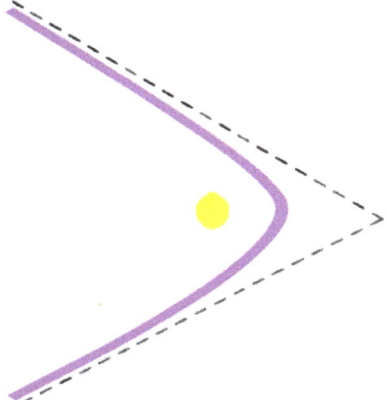

The shape reminds of a hyperbola branch from Chapter 3.2. Indeed, it is hyperbolic apart from the borderline case $\varepsilon=1$ which corresponds to a parabola. Hence, all gravity-driven paths are conic sections.

To confirm this, rewrite the solution as

$$r + \varepsilon r \cos\theta = r + \varepsilon x = r_0(1+\varepsilon).$$

The next steps depend on ε.

This demonstrates that $\varepsilon=1$ describes a parabola:

$$r + x = 2r_0 \rightarrow x^2 + y^2 = (2r_0 - x)^2 \rightarrow y^2 = 4r_0^2 - 4r_0 x \rightarrow -x = \frac{y^2}{4r_0} - r_0$$

This demonstrates that $\varepsilon>1$ describes a hyperbola:

Define $\delta = \varepsilon^2 - 1 > 0$ and $a = r_0/(\varepsilon - 1) > 0$.

$$r + \varepsilon x = -a\delta \rightarrow x^2 + y^2 = (\varepsilon x + a\delta)^2 \rightarrow (1-\varepsilon^2)x^2 - 2\varepsilon x a\delta + y^2 = a^2\delta^2$$

$$\rightarrow -\delta(x^2 + 2\varepsilon a x + \varepsilon^2 a^2) + y^2 = a^2\delta(\delta - \varepsilon^2) = -a^2\delta$$

$$\rightarrow \delta(x+\varepsilon a)^2 - y^2 = a^2\delta \rightarrow \frac{(x+\varepsilon a)^2}{a^2} - \frac{y^2}{\delta a^2} = 1$$

One practical application of hyperbolas involves space travel. Suppose we're heading for Neptune but don't have enough fuel to get there on our own. Fortunately, Jupiter isn't far out of our way. By steering close to Jupiter, we can hitchhike on its speed and gravity and get flung out the other side. This is called a gravity assist or slingshot.

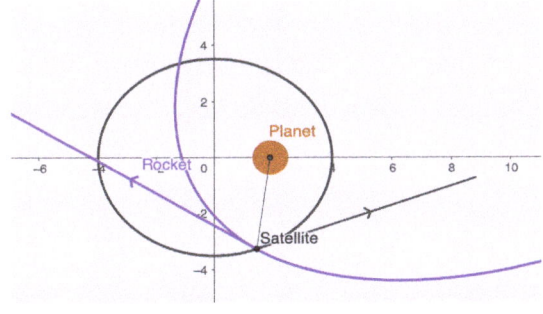

The GeoGebra activity "Rocket Launch from Satellite" lets you simulate gravity slingshots around a planet.

CHAPTER 7

7.3. (CO)SINE SIGNALS

Right Triangles

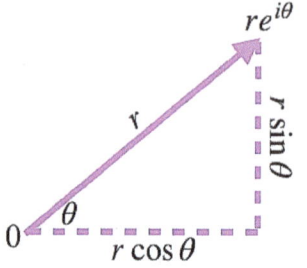
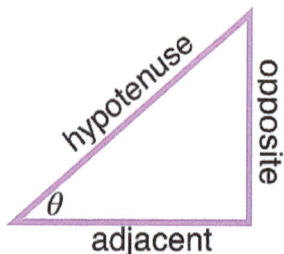

Above on the left is our now-familiar graph of $re^{i\theta}$ and its components. On the right is an ordinary right triangle with sides labeled relative to angle θ. Apart from markings, how do these two triangles compare?

While the hypotenuse on the right might not have length r, the corresponding angles match and so do the ratios of corresponding sides. Hence, we can define sine and cosine as

$$\cos\theta = \frac{\text{adjacent to }\theta}{\text{hypotenuse}}\ ;\quad \sin\theta = \frac{\text{opposite to }\theta}{\text{hypotenuse}}.$$

These are the standard definitions in trigonometry, which was invented nearly two millennia before calculus. They imply all other properties of sines and cosines. For example, while matching real and imaginary components of $(e^{i\theta})' = ie^{i\theta}$ quickly proves that $(\cos\theta)' = -\sin\theta$ and $(\sin\theta)' = \cos\theta$, the trigonometric proof below requires neither e nor i.

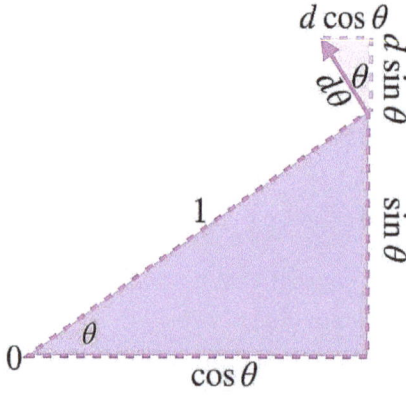

Travel in a unit circle from angle θ to $\theta + d\theta$ moves approximately $-d\cos\theta$ horizontally and $d\sin\theta$ vertically. The nudges form two sides of a right triangle with hypotenuse $d\theta$ tangent to the circle. Comparing the triangles confirms that

$$\frac{d\cos\theta}{d\theta} = -\sin\theta \ \text{ and } \ \frac{d\sin\theta}{d\theta} = \cos\theta.$$

Refraction

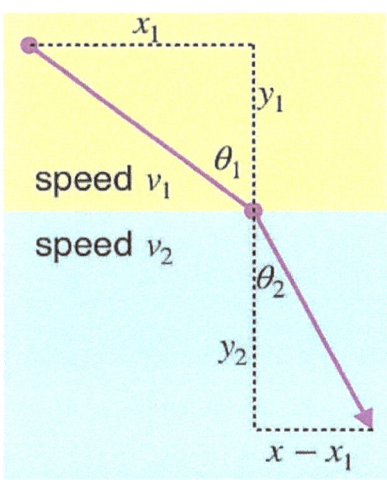

A lifeguard y_1 meters back from the shoreline spots a struggling swimmer y_2 meters offshore and x meters to the side. Since running at speed v_1 is faster than swimming at speed v_2, a straight path is not the quickest. What is the quickest path?

Clearly, the lifeguard should run and swim in straight lines, since straight is quickest for a given speed. For distance x_1 run laterally, the time spent running is $t_1 = \sqrt{x_1^2 + y_1^2}/v_1$ while the time spent swimming is $t_2 = \sqrt{(x-x_1)^2 + y_2^2}/v_2$. To minimize $t_1 + t_2$, differentiate with respect to x_1 and set $(t_1 + t_2)' = 0$, which requires

$$\frac{x_1}{v_1\sqrt{x_1^2 + y_1^2}} = \frac{x - x_1}{v_2\sqrt{(x-x_1)^2 + y_2^2}}.$$

Simplify to $\sin\theta_1/v_1 = \sin\theta_2/v_2$, where θ_1 and θ_2 denote the angles of motion relative to a line perpendicular to shore. The slower activity will cut a tighter angle. First-order Taylor approximations suggest setting the angles roughly proportional to the speeds.

Light generally behaves as if it too were seeking the quickest path to its destination. If we express light velocity as base speed c divided by a refractive index n, we obtain what is known as Snell's law of refraction:

$$n_1 \sin\theta_1 = n_2 \sin\theta_2.$$

For materials transparent to visible light, refractive indices rarely exceed 2. The index is 1.0003 for air, 1.333 for water, and $1.5-1.6$ for most glass. Diamond is exceptional with index 2.417.

CHAPTER 7

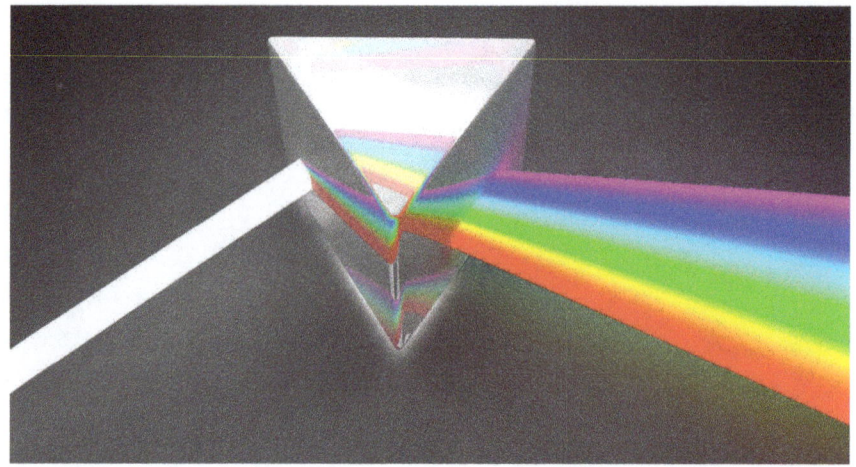

Refractive indices often vary with wavelength. Glass slows shorter-wavelength light more, which splits white light into a colored spectrum. Refraction of light through raindrops produces rainbows.

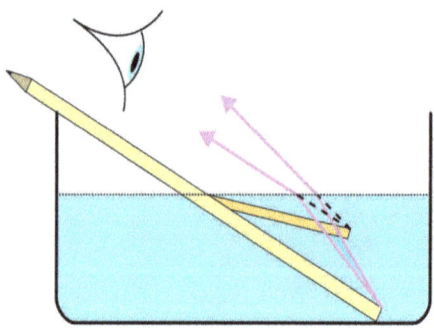

Refraction makes a pencil sticking into a bowl of water look kinked at the point of entry, with the segment below the surface appearing closer than it is. Through habit our brains presume that light travels straight and project the location wrong.

The view up from the water is even more curious. An angle of $\theta_1 = 90°$ flat to the water translates into $\theta_2 = 48.6°$ below. Hence the central $97°$ contains the whole vista above the surface, although with severe compression around the edges. A maximally wide-angle lens is known as a fisheye lens for this reason.

Internal Reflection

zThe low sine in Snell's Law cannot exceed $N = n_{low}/n_{high}$ as otherwise the high sine would exceed 1. The maximum or critical angle is the inverse sine or arcsine of N, written $\sin^{-1} N$ or $\arcsin N$. What happens if we look through the slow side of the interface beyond the critical angle? We can't. The only light we see is reflection, as the interface acts like a mirror.

> **Notation Alert**
>
> *For any trigonometric function* trig, trig^m *refers to its mth power, except for $m = -1$ which refers to its inverse* arctrig.

This underwater photo catches both the turtle and its reflection on the surface, but nothing above water.

High internal reflection is treasured. A diamond in air internally reflects any light steeper than $24.5°$. By cutting dozens of shallow facets with numerous symmetries, a skilled diamond cutter makes light scatter throughout the diamond and reflect back to the observer from many angles. Small shifts in the diamond cause abrupt shifts in brightness, which we see as sparkle.

CHAPTER 7

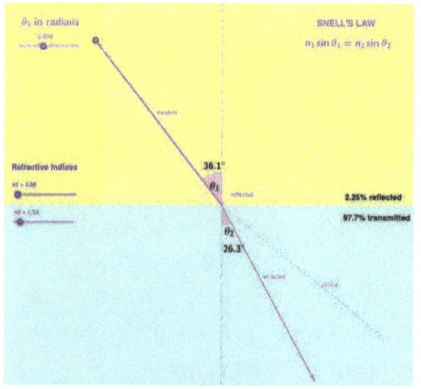

The GeoGebra activity 'Snell's Law" lets you explore refraction and critical angles. Adding realism, it allows light to be partially reflected even when most is refracted. This follows from conservation of total energy, although the derivations are too complex to pursue here.

Internal reflection is also key to the network of fiber optic cable threading the internet together. Optical fiber is transparent glass or plastic extruded as threads slightly thicker than human hair. By keeping each fiber very thin, light reflects internally over long distances with hardly any loss of power. Moreover, each fiber can carry millions of communications simultaneously, thanks to other applications of higher math.

Pendulums

Grandfather clocks use pendulums to tell time. A bob of mass m at the end of a thin rod of length L is suspended from a pivot and allowed to swing freely. Each swing turns the hand the same fraction of degree, which is calibrated to match the period T of the swing. However, the amplitude of the swing isn't constant. It might be high or low and might decay over time through tiny friction. How does the clock adjust for that?

The answer is: it doesn't. When the maximum swings are modest, say less than $30°$ on either side of vertical, T is relatively constant. We can confirm that using calculus.

This diagram displays the relevant forces. Gravity pulls down with force mg, which subdivides into a radial pull $mg\cos\theta$ and a transverse pull $mg\sin\theta$.

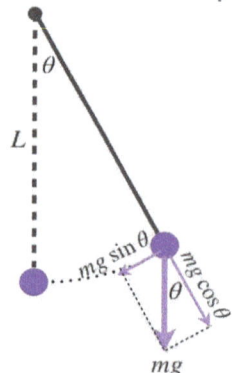

Since the rod is rigid, only the transverse pull affects motion. It induces acceleration $L\theta''$ in bob velocity $L\theta'$. Hence the driving equation is

$$\theta'' = -\frac{g}{L}\sin\theta.$$

This equation cannot be solved exactly using the functions we've met. However, modest swings justify the second-order Taylor approximation $\sin\theta \approx \theta$. This matches the equation for an ordinary spring, which from Chapter 5.4 has solution $\theta = \theta_0 \cos(t\sqrt{g/L})$ given a start from rest at angle θ_0. Since a full oscillation increases θ by 2π,

$$T \approx 2\pi\sqrt{L/g}.$$

For a rod of length 1 meter and $g \approx 9.81$ m/s², $T \approx 2.006$ seconds. This is not an accident, as the meter was originally proposed as the length of a pendulum with a half-period of one second.

Integrals of Powers of (Co)sines

Many neat tricks have been developed to integrate powers of sines and cosines Suppose the power on sine is an odd whole number $2m+1$. For $x = \cos\theta$, applying $\sin^2\theta = 1 - \cos^2\theta$ and $d\cos\theta = -\sin\theta\, d\theta$ shows that

$$\int_0^v \sin^{2m+1}\theta \cos^n\theta\, d\theta = -\int_1^{\cos v}(1-x^2)^m x^n dx.$$

Then expand the right-hand side into powers of x, integrate term by term, and convert back into sines and cosines. If the power on cosine is an odd whole number $2n+1$, a similar procedure for $x = \sin\theta$ yields

$$\int_0^v \sin^m\theta \cos^{2n+1}\theta\, d\theta = \int_0^{\sin v} x^m(1-x^2)^n dx$$

and proceed as before. When both sine and cosine have positive even powers, apply $\tau = 2\theta$, $\sin\tau = 2\sin\theta\cos\theta$ and $\cos\tau = \cos\tau = 2\cos^2\theta - 1$. For $n > m$ this yields

$$\int_0^v \sin^{2m}\theta \cos^{2n}\theta\, d\theta = \frac{1}{2^{n+1}}\int_0^{2v}\sin^m\tau(1+\cos\tau)^{n-m} d\tau.$$

For $m > n$, switch m with n, apply $\cos\tau = 1 - 2\sin^2\theta$ and change $1 + \cos\tau$ to $1 - \cos\tau$. Keep applying this method until odd powers of sine or cosine appear and then convert to powers of x.

To integrate $e^{p\theta} \sin^m \theta \cos^n \theta$, convert to complex exponentials using

$$\cos\theta = \tfrac{1}{2}\left(e^{i\theta} + e^{-i\theta}\right) \quad \text{and} \quad \sin\theta = -\tfrac{1}{2}i\left(e^{i\theta} - e^{-i\theta}\right).$$

When m and n are whole numbers or zero, substitution yields terms in $\exp((p+ik)\theta)$ for integers k ranging from $-m-n$ to $m+n$. Each term is easy to integrate; just divide it by $p+ik$. If desired, convert back to terms in real exponentials, sines and cosines.

To integrate $\theta^j e^{p\theta} \sin^m \theta \cos^n \theta$ for whole powers of θ, use integration by parts. The key procedure is recursive. Given any complex number c,

$$\int_0^\tau \theta^j e^{c\theta} d\theta = \frac{1}{c}\tau^j e^{c\tau} - \frac{j}{c}\int_0^\tau \theta^{j-1} e^{c\theta} d\theta.$$

While the previous methods convert sines and cosines into other expressions, sometimes it is useful to convert the roots of polynomials into sines and cosines. Consider the area under $\sqrt{1-x^2}$ from 0 to $z \le 1$. If we substitute $\sin\theta = x$, then $\cos\theta = \sqrt{1-x^2}$, $\cos\theta\, d\theta = dx$, and

$$\int_0^z \sqrt{1-x^2}\, dx = \int_0^{\theta(z)} \cos^2\theta\, d\theta = \int_0^{\theta(z)} \frac{1+\cos 2\theta}{2} d\theta$$

$$= \frac{\theta_z}{2} + \frac{\sin 2\theta_z}{4} = \frac{\theta_z + \sin\theta_z \cos\theta_z}{2} = \frac{\sin^{-1} z + z\sqrt{1-z^2}}{2}.$$

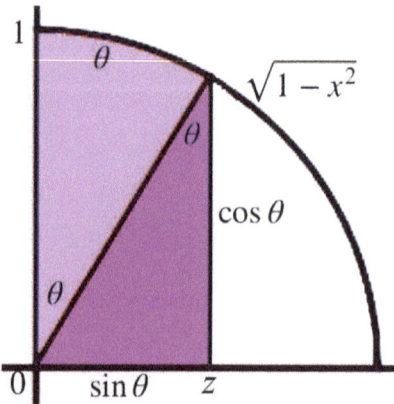

In words, the integral equals the sum of two areas: a θ radian slice of a unit circle and the triangle underneath. This is exactly what we want. The $\sin\theta = x$ substitution also shows that

$$\int \frac{dx}{\sqrt{1-x^2}} = \sin^{-1} x + C.$$

7.4 TANGENTS AND MORE

Tangents of Angles

Trigonometry relates angles to the ratio of two sides of a right triangle. Since any two sides determine the third, any single trigonometric function could potentially serve for all. Still, it is useful to know at least sine, cosine and their ratio:

$$tangent : \tan\theta = \frac{\sin\theta}{\cos\theta} = \frac{\text{opposite to }\theta}{\text{adjacent to }\theta}.$$

When the adjacent side extends to the right and the opposite side extends up, tangent equals the slope of the hypotenuse. The name is confusing, since it can also refer to a tangent line. To distinguish the meanings, remember that tan() is associated with an angle while a tangent line is associated with a function. To convert tan() back to cosine and sine, use the identities

$$\cos^2\theta = \frac{\cos^2\theta}{\cos^2\theta + \sin^2\theta} = \frac{1}{1+\tan^2\theta} \quad \text{and} \quad \sin^2\theta = \frac{\tan^2\theta}{1+\tan^2\theta}.$$

To calculate the derivative of $\tan\theta$, apply the product rule:

$$\frac{d\tan\theta}{d\theta} = \frac{1}{\cos\theta}\cdot\frac{d\sin\theta}{d\theta} + \sin\theta\cdot\frac{d(1/\cos\theta)}{d\theta}$$

$$= 1 + \sin\theta\cdot\frac{\sin\theta}{\cos^2\theta} = 1 + \tan^2\theta = \frac{1}{\cos^2\theta}.$$

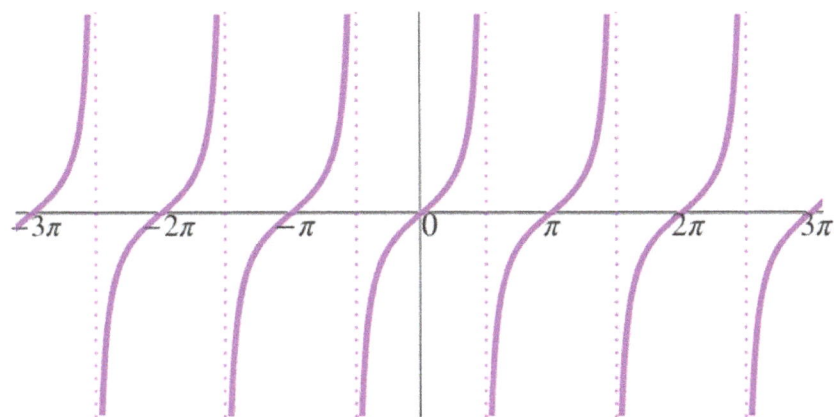

Hence $\tan\theta$ slopes upwards wherever the slope is defined, with minimum slope 1 and no finite maximum. For every integer k, $(k-\frac{1}{2})\pi$ to $(k+\frac{1}{2})\pi$ marks a strip where $\tan\theta$ grows from $-\infty$ to $+\infty$ and is not defined at the boundaries.

The area across tangent strips is undefined. The area under a single tangent strip is

$$\int_{\theta_1}^{\theta_2} \tan\theta\, d\theta = \int_{\theta_1}^{\theta_2} \frac{-d\cos\theta}{\cos\theta} = \ln\left(\frac{\cos\theta_1}{\cos\theta_2}\right).$$

The inverse tangent, written arctan or $\tan^{-1}$, denotes an angle associated with a given slope $x = \tan\theta$. Convention favors the angle closest to 0, which places it between $-\frac{1}{2}\pi$ and $+\frac{1}{2}\pi$. Since $(\tan\theta)' = \tan^2\theta + 1$,

$$\frac{d\tan^{-1}x}{dx} = \frac{1}{x^2+1}.$$

This relation will come in handy below.

Rational Polynomials

Rational polynomials are ratios of polynomials. Let us write these as

$$R(x) = \frac{P(x)}{Q(x)} = \frac{p_m x^m + p_{m-1} x^{m-1} \cdots + p_0}{x^n + q_{n-1} x^{n-1} + \cdots + q_0}$$

and examine the integral $\int_{x_1}^{x_2} R(x)\, dx$. When $m \geq n$, we can separate $R(x)$ into a polynomial of order $m-n$ plus a residual fraction with numerator of order less than n. The polynomial is easy to integrate: just substitute $x^{k+1}/(k+1)$ for every x^k. The big challenge is to integrate the residual fraction.

Warning: this gets messy. $R(x)$ flips between $-\infty$ and $+\infty$ wherever Q crosses the x-axis. Also, residual fractions approach zero at large x values but not as exact powers or exponentials. Fortunately, five methods cover all possibilities.

CIRCLE FUNCTIONS

1. When $n=1$, $\int_{x_1}^{x_2} \dfrac{p_0 \, dx}{x+q_0} = p_0 \ln \dfrac{x_2+q_0}{x_1+q_0}$, which is undefined when x_1 and x_2 lie on different sides of $-q_0$.

2. When $n=2$, define $a = \tfrac{1}{2} q_1$ and $b = \sqrt{|q_0 - a^2|}$. For $m=0$ and $q_0 \le a^2$, $Q(x) = (x+a)^2 - b^2$. If $b=0$, $R(x)$ integrates to $-p_0/(x+a)$. Otherwise rewrite $R(x)$ as $\dfrac{p_0}{2b}\left(\dfrac{1}{x+a-b} - \dfrac{1}{x+a+b}\right)$ and apply method 1.

3. When $n=2$, $m=0$, and $q_0 > a^2$, $Q(x) = (x+a)^2 + b^2$. Substitute $y = (x+a)/b$ and $dy = dx/b$ to obtain $\dfrac{p_0}{b} \int \dfrac{dy}{y^2+1} = \dfrac{p_0}{b} \tan^{-1} \dfrac{x+a}{b}$.

4. When $n=2$ and $m=1$, use $Q'(x) = 2x + q_1$ and $s = p_0 - ap_1$ to rewrite $\dfrac{P}{Q} = \tfrac{1}{2} p_1 \dfrac{Q'}{Q} + \dfrac{s}{Q}$. The first term integrates to $\tfrac{1}{2} p_1 \ln \dfrac{Q(x_2)}{Q(x_1)}$. Integrate $s/Q(x)$ using method 2 or 3.

5. When $n > 2$, factor Q into a product $Q_1 Q_2 \cdots Q_k$ of linear and quadratic terms. Then decompose R into

$$R(x) = \dfrac{P_1(x)}{Q_1(x)} + \dfrac{P_2(x)}{Q_2(x)} + \cdots + \dfrac{P_k(x)}{Q_k(x)},$$

where P_j is constant for linear Q_j and linear for quadratic Q_j. Then apply methods 1–4.

Similar methods serve to integrate any rational polynomial in sines and cosines. Note that

$$\cos 2\theta = \cos^2 \theta - \sin^2 \theta = \dfrac{\cos^2 \theta - \sin^2 \theta}{\cos^2 \theta + \sin^2 \theta} = \dfrac{1 - \tan^2 \theta}{1 + \tan^2 \theta}$$

$$\sin 2\theta = 2 \tan \theta \cos^2 \theta = \dfrac{2 \tan \theta}{1 + \tan^2 \theta}.$$

Hence, if we define $x = \tan \dfrac{\theta}{2}$, we can substitute $\dfrac{1-x^2}{1+x^2}$ for $\cos \theta$, $\dfrac{2x}{1+x^2}$ for $\sin \theta$, and $\dfrac{2\,dx}{1+x^2}$ for $d\theta$. Simplify into the form $P(x)/Q(x)$. Tedious? Yes But computers can handle most of the drudge.

CHAPTER 7

Trigonometric Reciprocals

There are special names for the reciprocals of sine, cosine, and tangent:

secta $:\sec\theta = \dfrac{1}{\cos\theta}$

cosecta $:\csc\theta = \dfrac{1}{\sin\theta}$,

cotg h $:\cot\theta = \dfrac{\cos\theta}{\sin\theta}$

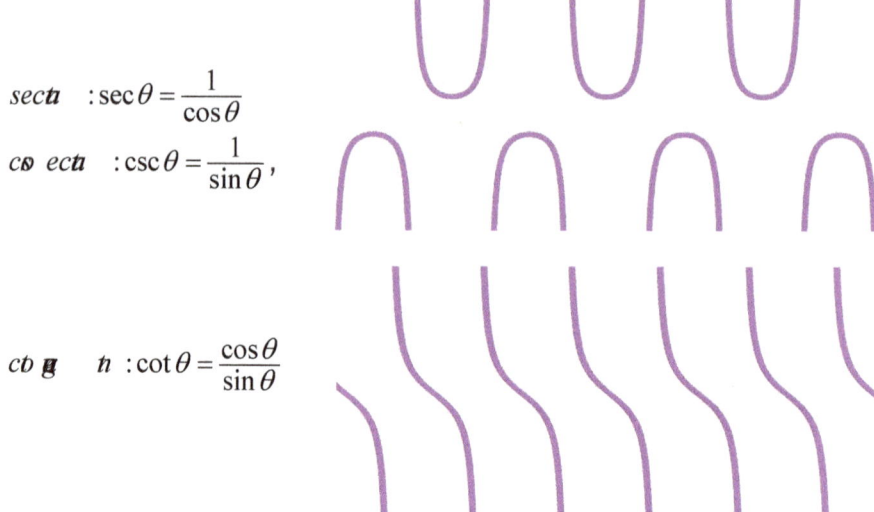

These are poor choices for names as "secant" overlaps with a chord between two points on a curve and the "co-" prefixes lack the common naming principle they suggest. They serve mainly to vex students taking trigonometry exams. Still, good learners should be aware of various connections between the concepts and be able to re-derive as needed.

The inverses crop up mainly in the following integrals, the first of which we have already seen:

$$\int \dfrac{dx}{x^2+1} = \tan^{-1}x + C = -\cot^{-1}x + C$$

$$\int \dfrac{dx}{|x|\sqrt{x^2-1}} = \sec^{-1}x + C = -\csc^{-1}x + C.$$

$(\sec\theta)' = \dfrac{-1}{\cos^2\theta}(\cos\theta)' = \dfrac{\sqrt{1-\cos^2\theta}}{\cos^2\theta}$

$= \dfrac{1}{\cos\theta} \cdot \sqrt{\dfrac{1-\cos^2\theta}{\cos^2\theta}} = \sec\theta\sqrt{\sec^2\theta - 1}$

To confirm the second integral, note that the derivative of $\theta = \sec^{-1}x$ is the reciprocal of the expression to the left.

Logarithmic Spirals

Let's close with a soaring application. Due to the shape of their heads and location of eye sockets, hawks see better diagonally than straight ahead. To keep better watch, they approach their prey at best viewing angle and spiral in. How can we describe the spiral mathematically?

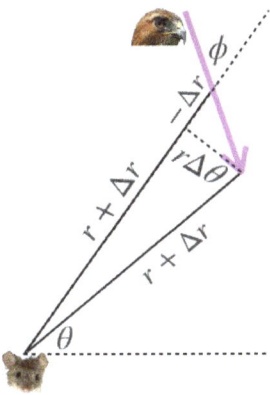

The simplest model treats the hawk's line of sight to prey as a radial vector of length r and angle θ. In a short time Δt, the hawk flies $\Delta r \approx r'\Delta t$ closer and about $r\Delta\theta \approx r\theta'\Delta t$ off to the side. If ϕ ("phi") denotes the hawk's pitch or angle off target, $\cot\phi \approx -\Delta r/(r\Delta\theta)$. In the limit of tiny Δt,

$$\cot\phi = \frac{-1}{r} \cdot \frac{dr}{d\theta} = -\frac{d\ln r}{d\theta}.$$

The solution is $r = b \cdot \exp(-\cot\phi \cdot \theta)$, where b is set to match the hawk's initial position. This is known as a logarithmic spiral.

A log spiral is self-similar, meaning that each part looks identical apart from scale. Every circuit shrinks distance by a factor of $\exp(2\pi|\cot\phi|)$. The case $\cot\phi = 0$ corresponds to a pitch of 90° with hawk in constant circle. Reducing ϕ makes the circuit shrink faster. At $\phi = 0$ the hawk heads straight for the prey. For $\phi < 0$, as pictured here, the hawk circles clockwise rather than counterclockwise.

In principle, each spiral wraps infinitely many times around the prey. Nevertheless, the hawk needn't starve along the way. For every infinitesimal length dL that the hawk flies, r shrinks by $\cos\phi \, dL$, so

$$dL = \frac{-dr}{\cos\phi} = -\sec\phi \, dr.$$

Hence total distance $L = r_0 \sec\phi$ is finite. Empirically the hawk's pitch is about $40°$, making flight distance about 30% longer than straight, but with flight time roughly the same thanks to reduced drag.

Insects approach light sources in log spirals. A constant pitch keeps their path nearly straight until close to the source. Slanted growth in plants and animals, trailing clouds in hurricanes, and trailing stars in rotating galaxies approximate log spirals too.

The GeoGebra activity "Logarithmic Spiral" lets you make your own.

With that, dear reader, our games end. But if you keep playing with calculus, you'll find plenty of opportunities to use it. What a wonderful tool for analyzing our wondrous world!

Credits

Note: All Wikimedia images are licensed under the Creative Commons Attribution-ShareAlike 4.0 International License, (www.creativecommons.org/licenses/by-sa/4.0), the previous 3.0 license, or GFDL (www.gnu.org/copyleft/fdl.html)]. They are available at commons.wikimedia.org/wiki/File: followed by "NameTag.xyz" given below.

Many of the images cited below, whether from Wikimedia or other sources, have been cropped or recolored from the originals.

All GeoGebra activities mentioned in the text can be accessed at www.geogebra.org using the activity name and/or author tag "kosband". The original author is cited in each online heading.

Chapter 0

p 2, small pebbles: Pseudopanax, Wikimedia, Small_gray_pebbles.jpg

p 4, counting bears: Learning Resources, Bear Counters Set

p 4, pencils: Alexandra Koch, pixabay.com/illustrations/colour-pencils-pens-paint-4214617/

p 4, inflatable ball: pixabay.com/vectors/beach-ball-ball-inflatable-beach-575425/

p 4, geometric orb: Gordon Johnson, pixabay.com/vectors/geometric-orb-ball-sphere-globe-4663871/

p 4, "2": pixabay.com/vectors/number-2-digit-figure-cipher-150791/

CREDITS

p 8, bubbles: Soap Bubbles on Blue Sky, HP_Photo/stock.adobe.com

p 8, water fountain: GuidoB, Wikimedia, ParabolicWaterTrajectory.jpg

p 8, exponential growth of gold: © Bojan Dzodan, Dreamstime.com

p 9, roller coaster platform: Jo Jakeman, flickr.com/photos/jojakeman/3512781378

p 9, diamond: Large Beautiful Diamond, AlexMas/stock.adobe.com

p 10, baffled bug: pixabay.com/en/bug-germ-virus-viruses-alien-268531/

p 11, find x: Wade Clarke, Ocular Trauma, issue 185, July 13, 2005, ocular-trauma.net/gallery.shtml?a

p 11, boy and dog: © Everett Collection, Dreamstime.com

Chapter 1

p 26, bubbles: Soap bubbles on blue sky, HP_Photo/stock.adobe.com

p 28, fruit: James Peale, Still Life with Fruit, Wikimedia, Still_Life_with_Fruit_by_James_Peale,_c._1821,_De_Young_Museum.jpg

p 28, pigs: Alvesgaspar, Wikimedia, Pot-bellied_pigs_in_Lisbon_Zoo_2008.jpg

p 29, hand: Iris Vallejo, pixabay.com/photos/hand-elderly-woman-wrinkles-351277/

p 30, Rumplestiltskin: Anne Anderson, Wikimedia, Rumplestiltskin_-_Anne_Anderson.jpg

p 31, skull and crossbones: wpclipart.com/signs_symbol/skull/skull_and_crossbones_large.png

p 33, vortex: pixabay.com/en/whirlpool-funnel-tornado-vortex-30390/

Chapter 2

p 41, inscribed polygon: KSmrq, Wikimedia, Archimedes_circle_area_proof_-_inscribed_polygons.png

CREDITS

p 41, circumscribed polygons: KSMrq, Wikimedia, Archimedes_circle_area_proof_-_circumscribed_polygons.png

p 42, red onion slice: © Robyn Mackenzie, Dreamstime.com

p 49, lower sum and upper sum: KSmrq, Wikimedia, Riemann_sum_convergence.png

p 50, definite integral: KSmrq, Wikimedia, Integral_example.svg

p 55, ice cube: openclipart.org/people/lekamie/ice-cube.svg

p 56, wax pyramid: meltingcandle-wax.blogspot.com

p 56, diagonals inside hollow cube: Robert Webb (Stella software), software3d.com/Stella.php, Wikimedia, Cubic_pyramid.png

p 56, cube decomposed into square pyramids: Six Identical Pyramids from Cube, 3dwarehouse.sketchup.com

p 58, Earth layers: Earth layers, piai/stock.adobe.com

p 59, pyramid section inside sphere: Utah Electronic High School, share.ehs.uen.org/sites/default/files/images/volumesphere.png

p 60, spherical ring: Ag2gaeh, Wikimedia, Kugel-ring-s.svg

Chapter 3

p 66, camel: J. Patrick Fischer, Wikimedia, 2011_Trampeltier_1528.jpg

p 72, radio telescopes: Very Large Array, Joe Gough/adobe.stock.com

p 75, conic sections: Magister Mathematicae, Wikimedia, Conic_Sections.svg

Chapter 4

p 93, girl dropping ball: Center for Particle Astrophysics, Berkeley Cosmology Group, cfpa.berkeley.edu/Education/ISTAT/cfpa/speed/speedweb.html

p 95, Earth's gravitational pull: Gravity © Dannyphoto8, Dreamstime.com

CREDITS

p 97, escape velocities for solar system: hyperphysics.phy-astr.gsu.edu/hbase/Solar/soldata 2.html

p 97, Celtic hammer throw: Frank R. Stockton, Round-about Rambles in Lands of Fact and Fancy, gutenberg.org/files/17582/17582-h/17582-h.htm #GYMNASTICS

p 101, tidal forces on Earth: Krishnavedala, Wikimedia, Field_tidal.svg and overlaid on globe from Jcpag2012, Wikimedia, 3D_Earth.png

p 102, centrifuge: NASA, nasa.gov/centers/ames/multimedia/images/2006/20gcentrifuge.html

p 104, sprinters around bend: Zen, Wikimedia, 200metres Helsinki-2005.jpg

p 104, speed skaters: Fiona van Doorn, Wikimedia, Annita-van-doorn-en-liesbeth-mau-asam-1300458205.jpg

p 104, motocross racers: Robert Scoble, Wikimedia, MotoGP_final_race.jpg

p 104, bicycle racers: Gsi, Wikimedia, Robbie_McEwen_2007_Bay_Cycling_Classic_2.jpg

p 105, auto race: Willowbrook Hotels, Chicagoland Motor Speedway, flickr.com/photos/29573712@N08/2757563789/

p 106, jet turning: Deeday-UK, Wikimedia, Banked_turn.png

p 107, roller coaster structure: Wikimedia, Incredible_Hulk_Coaster.jpg

Chapter 5

p 132. Hooke's Law springs: Svjo, Wikimedia, Hookes-law-springs.png

Chapter 6

p 138, log scales for multiplication: Jakob.scholbach, Wikimedia, Slide_rule_example2_with_labels.svg

p 138, slide rule: Adrian Tync, Wikimedia, Skala_slide_rule.jpg

CREDITS

p 139, circular slide rule: author's photo

p 139, decibel scale: Decibel, desdemona72/stock.adobe.com

p 141, pH scale: PH scale, © Alain Lacroix, Dreamstime.com

p 142, electromagnetic spectrum: Philip Ronan, Wikimedia, EM_spectrum.svg

p 145, exponential growth of gold: © Bojan Dzodan, Dreamstime.com

p 149, rocket propulsion: pixabay.com/vectors/rocket-missile-lift-off-start-fire-146104/

p 154: US debt to GDP history, Congressional Budget Office, The 2023 Long-term Budget Outlook, Figure 1.1, Deficits and Debt, www.cbo.gov/system/files/2023-06/59014-LTBO.pdf.

p 159, Population pyramid for richest countries: data from Most Developed Countries in 2019, US Census Bureau, International Data Base, census.gov/data-tools/demo/idb/informationGateway.php

p 159, Population pyramid for poorest countries: data from Least Developed Countries in 2019, US Census Bureau, International Data Base, census.gov/data-tools/demo/idb/informationGateway.php

p 161, oblong Lotka-Volterra orbits: Wiso, Wikimedia, Lotka-Volterra.svg

Chapter 7

p 163, complex number plane: Polar grid, pyty/stock.adobe.com

p 168, polar coordinate grid: Henjansson, Wikimedia, Logpolargrid.jpg

p 170, Kepler's Laws: Hankwang, Wikimedia, Kepler_laws_diagram.svg

p 178, prism spectrum: Prism light spectrum dispersion, Alexander Mitiuc/stock.adobe.com

p 178, refraction of pencil in water: derivative work by Theresa_knott of Gregors, Wikimedia, Pencil_in_a_bowl_of_water.svg

p 178, fisheye view of New York skyline: Peter Wieden, Wikimedia,

CREDITS

WFlatiron_fishView_ideal.jpg

p 179, underwater reflection of turtle: Brocken Inaglory, Wikimedia, Total_internal_reflection_of_Chelonia_mydas.jpg

p 179, uncut diamond: Eurico Zimbres, Wikimedia, DiamanteEZ.jpg

p 179, diamond: Large Beautiful Diamond, AlexMas/stock.adobe.com

p 180, pendulum: Charles H. Henderson and John F. Woodhull, Wikimedia, Grandfather_clock_pendulum.png

p 187, hawk head: Scott Campbell, Wikimedia, Northern-Red-Tailed-Hawk.jpg

p 187, mouse head: George Shuklin, Wikimedia, %D0%9C%D1%8B%D1%88%D1%8C_2.jpg

p 188, spiraling hurricane: NASA, rapidfire.sci.gsfc.nasa.gov/gallery/?20102560913

p 188, spiral galaxy: NASA and European Space Agency, antwrp.gsfc.nasa.gov/apod/ap050428.html

p 188, cutaway view of nautilus shell: Chris 73, Wikimedia, Nautilus-Cutaway-LogarithmicSpiral.jpg

Index

A
accompaniments, 12
acidity, 140-141
addition rule, 76-77
air drag, 151-152
airplanes, 105
aliens, 95-96
ambrosia, 122
angle, 23-24, 39, 68-69, 71-74, 94-95, 104, 165-169, 176-184,187-188
antilogarithm, 144
conceptual approach, 9-10
arc length, 166, 169, 187
artillery, 93-94

B
ball, 4, 43-45, 47, 52, 68, 70, 72, 93-94, 97,128
banked turns, 104-106
Bézier curves, 109-110
binomial theorem, 127-128
bubbles, 27

C
cake slices, 122-123
calculators, 111-112
centrifuges, 102-103

chain rule, 77-81, 85-86, 98, 106, 109, 133, 146, 149, 151, 160, 171
circles, 41-42, 57, 106-108
circular motion, 68, 98-99, 169-170, 173
Clarke orbit, 100
clothoid, 108
complex exponential, 165-169, 171, 176, 182
complex numbers, 162-169, 171, 180, 182
complex roots, 167
computer programming, 7
concavity, 90
continuous, 49-55, 64, 67, 91, 106, 119, 145, 147
cones, 57-58
constant relative growth, 129
continuous compounding, 145
converge, 33-34, 49, 83, 86, 113-114, 124-127, 129, 131, 144, 146, 160
convexity, 89-90
cooling, 150-151
Cornu spiral, 108
cosine, 144, 167-168, 176, 181-186
cubes, 55-56

curvature, 106-110
curve, 5-6, 8, 13, 48, 53, 64, 67-68, 83-84, 87-88, 90, 104-106, 108-110, 130, 158, 169, 175, 186

D
damped oscillations, 155-157, 160
debt bubbles, 154
decibels, 139
degrees of smoothness, 91-92
delta-x convention, 31-32
de Moivre's formula, 167
differential equations, 134
differentials, 35-36, 140, 163
displacement, 47-48
distance, 13, 18, 43-50, 52-53, 57, 70-71, 74, 89, 92, 96, 99, 101, 107-108, 138, 141, 167-168, 172, 177, 180, 187-188
double integrals, 91
doubling time, 147-148

E
Earth, 8, 58, 93, 95-97, 99-102, 149, 170
earthquakes, 140
electromagnetic radiation, 141-142
ellipses, 70-75, 79, 171-172
escape velocity, 97, 149
Euler's formula, 165
Euler's identity, 166
existence of derivatives, 64
exponentials, 130, 146-147, 150, 154-159, 165, 172, 182, 184

F
falling around, 99-100
fetch, 43-45
finding zeros, 115
force, 87, 92, 95-108, 132, 134, 141, 149, 151, 158

G
geometric series, 122-124
grassy slopes, 61-63
gravity, 92-105, 132, 149, 171, 174-175

H
half-life, 147-148
hammer throw, 97-98
hawk in flight, 187-188
higher-order approximation, 117-118
hyperbolas, 73-75, 79, 175

I
imaginary numbers, 162-167
induction, 82-83
inflection points, 90-91
integral notation, 51-52
integrals as functions, 50-51
integration by parts, 81
interest rate, 145-146
irrational powers, 85-86
isosceles insights, 22-23

J
jerks, 107-108
jumps, 52-53, 64, 66

K
Kepler's Laws, 99-100, 170-174

L

L'Hôpital's rule, 135
light, 141-142
limits, 32-36, 38, 42, 46, 51, 56-59, 69, 72, 79, 86, 91, 123, 125, 127, 135, 145, 151, 166, 187
logarithm, 112, 131-132, 136-139, 142-144, 187-188
logarithmic spirals, 187-188
logistic growth, 157-158
Lotka-Volterra model, 160-161
loudness, 139

M

Maclaurin series, 118, 129, 131, 133, 165
maximum, 16, 25, 48, 50-51, 54, 60, 65-67, 70, 88-91, 95, 99, 101, 103, 118-119, 149, 155, 179-180
Mean Value Theorem, 118-120
minimum, 48, 51, 54, 66-67, 88-91, 99, 118-119, 184
monotone sequence, 124-127
moon, 101
multiplication rule, 79-81
multiplication, 37-39
myths, 1-7

N

negative powers, 83-84
Newton, 7, 13, 63, 89, 95, 98, 100, 129, 132, 136, 149-150, 155, 171
Newton-Raphson method, 116-117

notation, 4, 6, 9, 14, 30-31, 35-36, 51, 63, 76, 89, 118, 120, 179
nudges, 15-25, 27, 29, 31, 33, 35, 37, 39, 41, 43, 45, 47, 49, 51, 53-55, 57, 59
numerical estimation, 120-122

O

Oberth effect, 97
operators, 30
orbit, 99-103, 160-161, 170-175
osculating curve, 106-107
outline of book, 12-14
overdamping, 156

P

parabolas, 48, 55-56, 72-73, 75, 94, 117, 162, 175
Pascal's triangle, 127
pebble, 1-2, 12, 17-19
pendulum, 180-181
perpetuity, 154
pH scale, 140-141
phase shift, 167
playpen design, 15-17, 87-88
plumpness, 28
polar coordinates, 168
polynomial, 13-14, 86, 117-118, 120, 124, 129-135, 182-185
population pyramid, 158-159
predators and prey, 160-161
prerequisites, 11
present value, 152-154
Punch and Judy, 122-123, 136-137
pyramids, 55-59, 128, 159-160

R

radian, 166, 168, 182
radioactivity, 147-148
radius, 19, 27, 41-42, 57-60, 68-69, 96-100, 105-107, 127, 169-170, 172, 188
ratio tests, 125-126
rational polynomials, 184-185
rational powers, 84-85
reflection, 21, 32, 71-74, 87, 114, 179-180
refraction, 176-178
regularity, 23-25
relative growth rates, 142-143
Richter scale, 140
Riemann sums, 49-53, 121, 131, 137, 142
rockets, 97, 100, 148-149, 175
Rolle's theorem, 67, 118
roller coasters, 107-108
round design, 25-26
rotation, 38, 41, 70, 74-75, 83, 98, 100-104, 164, 167-169, 173, 188
rule of 70, 147
Rumpelstiltskin, 30

S

scale, 17, 57, 112, 138-141, 163, 165, 187
second derivative, 89, 92, 95, 108, 117
Simpson's rule, 121-122
sine, 144, 167-168, 181-186
sledding, 45-47
slice, 22-23, 40, 42, 57, 59-60, 69, 121-123, 182
slide rules, 138-139
Snell's Law, 176-178
spheres, 26-27, 57-60, 69, 79
spherical ring, 59-60
spline, 108-110
spring, 6, 8, 13, 132-134, 155-157, 173, 181
square root of -1, 162-167, 176
square roots, 111-114
stationary orbit, 100
strange staircases, 49-50
Sun, 8, 26, 96, 99-102, 104, 170-171, 174-175
surface 6, 27-28, 58-59, 69, 87, 96-97, 100, 102, 151

T

tangent lines, 61-74, 98, 102, 106, 115-119, 176
tangents of angles, 183-186
Taylor series, 117-135, 142-146, 166, 181
tennis solitaire, 68-69
tides, 101
top of the hill, 19-20
triangles, 20-23, 39-42, 46-49
trigonometric reciprocals, 186
types of convergence, 124-125

U

underdamping, 155
unit rule, 76

V

vision of future, 14
volume, 5-6, 27-28, 37, 55-60, 79, 87

W

wedding bands, 59-61
wrinkles, 29

www.ingramcontent.com/pod-product-compliance
Lightning Source LLC
Chambersburg PA
CBHW040759150426
42811CB00055B/1073

This book belongs to

www.ingramcontent.com/pod-product-compliance
Lightning Source LLC
Chambersburg PA
CBHW040801150426
42811CB00056B/1120